The Encyclopedia of High Fidelity

ACOUSTICS

The Encyclopedia of High Fidelity

Edited by John Borwick

ACOUSTICS
G. W. Mackenzie

AMPLIFIERS
H. Lewis York

DISC RECORDING AND REPRODUCTION
P. J. Guy

TAPE RECORDING AND REPRODUCTION
A. A. McWilliams

RADIO RECEPTION
H. Henderson

LOUDSPEAKERS
E. J. Jordan

ACOUSTICS

G. W. Mackenzie

Focal Press

New York & London

FIRST PUBLISHED 1964

PRINTED IN GREAT BRITAIN BY
*Richard Clay and Company, Ltd.,
Bungay, Suffolk*

CONTENTS

EDITOR'S INTRODUCTION

High Fidelity—the recording and reproduction of sound with maximum faithfulness to the original—has been in existence for only a decade or two. Yet in that time it has become an absorbing hobby for thousands of technical and musical enthusiasts. In countless homes it provides the key to unlimited musical enjoyment in the form of today's high-quality phonograph records, tape and FM broadcasting.

During the same period high fidelity has risen in importance as a subject of study at technical institutes and universities, where it is variously listed as Audio, Electronic, Broadcast and Telecommunication Engineering.

The decision to produce this series of books on the techniques of sound reproduction in such a way as to make them add up to a comprehensive Encyclopedia of High Fidelity in six volumes arose from the following considerations:

1. To treat this wide subject adequately in a single book and at a level suited to the needs of both the technical student and the serious amateur would require a work of unmanageable proportions.

2. The practice of assembling high-fidelity equipment in separate components conveniently allows the student and the amateur to study or work separately at these six aspects—Acoustics, Amplifiers, Loudspeakers, Disc, Tape and Radio.

3. In this age of specialisation, we were presented with the seeming paradox that six expert specialist authors were easier to find and brief than one single expert on all high-fidelity techniques.

Accordingly, and before one word was written, the Editor was able to hold a series of meetings with the authors, each of whom is an expert in his particular field. This procedure has insured that the technical level is uniform throughout the series, and that the volumes dovetail to provide complete coverage at the same time that they take their place in the literature as individual works in their own right.

Special attention has been given to terminology. Each book includes its own Glossary of Terms. Anyone possessing the complete series has access to a sectionalised dictionary and reference to the whole subject of sound recording and reproduction techniques.

JOHN BORWICK

PREFACE

THE aim of this book has been to select those parts of the very wide field of acoustics which are of interest to the amateur hi-fi enthusiast. In this age of tape recorders, disc reproducers, F.M. tuners, etc., many users are interested in the physics of sound—what sound waves are, how they are produced, how sound is picked up by microphones and how room acoustics affect the quality of sound.

Without going too deeply into the mathematics of the subject, the book describes and explains the important features of all sound sources, rooms and enclosures. This volume therefore acts as a general companion to the other books in this series devoted to specific topics. Microphones are treated in some detail since these play an important part in acoustic measurements and, of course, are now being used by thousands of tape recorder owners.

The author wishes to thank the following manufacturers for supplying and giving permission to publish information on their products:

Messrs. Standard Telephones and Cables Ltd., A.K.G. Vienna, Dawe Instruments Ltd., Cosmocord, Reslosound, R.C.A., Grampian, Sennheiser Electronics, Tannoy, Pamphonic Reproducers.

G. W. MACKENZIE

HOW SOUND WORKS

THE source of any sound is a vibrating body. This body can take one of several forms, for example the vocal cords by which we speak, the string on a violin, the air column in a clarinet, the cone of a loudspeaker.

The vibrating body causes pressure variations in the air surrounding it, and these variations pass through the air as a wave motion. This can affect our ears to produce the sensation of hearing or be picked up by a microphone which converts the pressure variations into voltage variations. These can then be amplified, for broadcasting or recording.

We see, therefore, that there are three essentials in acoustics—a vibrating body, a medium through which the sound wave can travel and some form of receiver. It should be noted that, like the vibrating body, the medium can be one of several forms. Air was mentioned first as it is the one we are normally accustomed to but all materials transmit sound waves to some degree, examples being water, metal and wood.

Creation of a Sound Wave

A medium can transmit waves because it possesses elasticity. This is the property whereby certain materials tend to return to their original shape or volume when the force causing a change is removed. A simple everyday example is rubber. If a piece of rubber is stretched and then released it will return to its original length. By contrast, a piece of lead will show little or no evidence of elasticity. Steel is another material which exhibits elasticity. Within limits, it returns to its original shape when a stretching force is removed.

11

A medium such as gas also exhibits elasticity, but in a different way. If a layer of a gas, such as air, has its volume changed, it returns to its original volume when the force causing the change is removed. It is this property of air which allows it to transmit the vibrations of a sounding body as a wave motion.

Taking a vibrating reed as a source, let us see how the wave is built up. Fig. 1.1 shows how the reed vibrates about its rest position.

FIGURE 1.1
A reed is chosen as an example of a simple source of sound. It vibrates about its rest position first to one side A and then to the other side B.

It moves first to A then back through its rest position to the other side B, reverses and moves to A again. Then the cycle is repeated over again.

When the reed moves to A, the volume conditions of the air surrounding the reed must be altered. The air particles in front of the reed will be pushed closer together and will thus be compressed. The particles behind it will be able to spread out to fill a bigger volume, and the air is said to be rarefied. Due to the elasticity of the air, these changes from normal conditions will affect further layers so that a compression will travel outwards from one side of the reed, and a rarefaction from the other.

When the reed moves over from A to B, the conditions are reversed. On side B there will now appear a compression, while on A there will be a rarefaction. With the reed vibrating continuously, alternate compressions and rarefactions are produced on both sides of the reed, and it is these variations in air pressure which constitute the sound wave.

Fig. 1.2 shows how the variations in pressure are superimposed on the standing atmospheric pressure. The size of the variations has been grossly exaggerated in the diagram. The standing atmospheric pressure is approximately 1,000,000 units per square centimeter, whereas pressure variations due to audible sound waves lie between 0·0002 units and 1,000 units per square centimeter (the unit used is the dyne, which is the force which will accelerate a mass of 1 grm by 1 centimetre per second per second).

12

Another way to look at a sound wave is to think of the air particles as being spheres which, under normal conditions, are distributed evenly throughout space as shown in Fig. 1.3a.

If we now create a sound wave, the particles will vibrate about their normal position. Suppose that we can freeze the air instantaneously and study the way the particles have moved. At the instant shown in Fig. 1.3b, particle 1 is at its rest position, 2 has moved forward slightly, 3 is farther forward, 4 at the farthest point from the rest position, 5 at the same distance as 3, 6 the same as 3, and 7 is at its rest position. Particles 8, 9, 10, 11, 12, have also moved from their rest positions, but in the opposite direction.

Such a wave motion, in which the particles vibrate along the direction in which the wave is travelling, is termed a *longitudinal* wave. In another type of wave, the particles of the medium vibrate at right angles to the direction in which the wave is travelling. The motion is then said to produce a *transverse* wave. Transverse waves occur on vibrating strings and on the surface of water, to take only two examples.

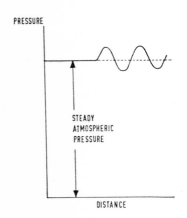

PRESSURE

STEADY
ATMOSPHERIC
PRESSURE

DISTANCE

FIGURE 1.2
When a vibrating body causes displacement of the air particles surrounding it, the air pressure varies. This pressure variation is very small and in the sketch has been magnified for ease of illustration.

There are two very important points to be noted about all types of wave motion:

1. Energy is propagated.
2. There is no displacement of the medium as a whole.

The first point is obvious, as energy must be available to cause vibrations in the receiver. The second point implies that the particles of the medium simply vibrate about their normal position of rest, as was shown in Fig. 1.3.

13

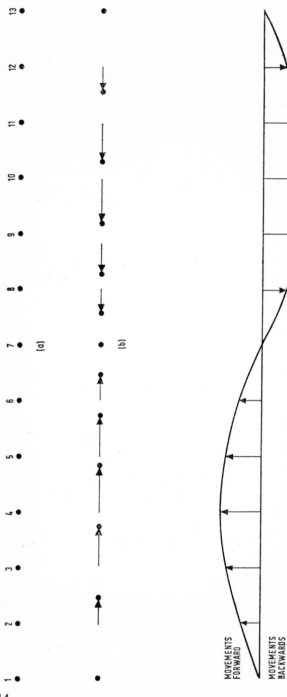

FIGURE 1.3

Air particles are represented by dots 1 to 13. (a) Normally we can assume they are spaced out evenly. (b) Under the influence of the sound waves radiated by a vibrating body, the relative positions of the particles are altered. (c) The displacements can be re-drawn to give the more usual 'wave'.

MOVEMENTS FORWARD

MOVEMENTS BACKWARDS

14

Representation of a Sound Wave

To represent waves in diagram form, one usually draws a transverse wave, as it is much easier to visualise. So that the longitudinal wave motion of a sound wave can be shown in a simple sketch, the convention used is to show forward movements above, and backward movements below a reference line which corresponds to the position of rest. This is shown in Fig. 1.3c. By joining the tips of the arrows together a normal transverse type of wave drawing is obtained.

Fig. 1.3 shows how the air particles are displaced by a sound wave. From this graph we can obtain another diagram which shows how the pressure varies. Fig. 1.4a, b and c are the same as Fig. 1.3, except that more particles are shown. It we look at particle 7 in Fig. 1.4b, we find it is at its rest position and the particles on either side of it, i.e. 6 and 8, have moved towards it. Exactly the same conditions apply at particle 19. But at particle 13 the particles on either side have moved away, and exactly the same conditions apply at particles 1 and 25. Obviously the pressure must be high around particles 7 and 19 and low around particles 1, 13 and 25. In fact 7 and 19 are the centres of *compressions*, and 1, 13 and 25 are the centres of *rarefactions*. The pressure on particles 4, 10, 16 and 22 must be normal, as the particles on either side of them have moved the same distance from their rest positions. If we now draw up all of this information about the pressure at different points in space, we obtain Fig. 1.4d. This shows the pressure wave which results from the displacement wave of Fig. 1.4c.

Figs. 1.3 and 1.4, show how the displacement and pressure vary in space or, to be more precise, how they vary with distance from the source. But another aspect is how they vary with time. A simple way of distinguishing between the ideas of distance and time is to appreciate that when we are considering distance we are looking at the sound wave as a *whole* at one particular instant of time. When we consider time, we are looking at one particular particle. For example, look at particle 4 in Fig. 1.4b. At the time chosen for examining the wave as a whole, particle 4 is at its maximum displacement in the forward direction. A fraction of a second later it must reverse its direction (remember that the air as a whole does not move forward). Later in time, the particle must reach its rest position and pass through it until it reaches a maximum displacement in the backward direction. Then it reverses and moves forward again, passing through the rest position once again until it reaches maximum displacement in the forward direction—which was our starting point. This whole chain or cycle of events is shown in Fig. 1.5a.

15

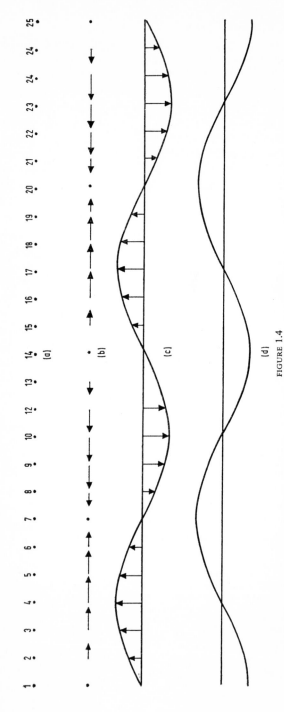

FIGURE 1.4

Using a similar diagram to Fig. 1.3 we can see how the displacement of the air particles causes a pressure wave to be set up. (a), (b) and (c) are as in Fig. 1.3, except that more particles have been included. (d) Because of the air vibrations, the pressure at different points in space varies in a wave-like manner similar to the displacement.

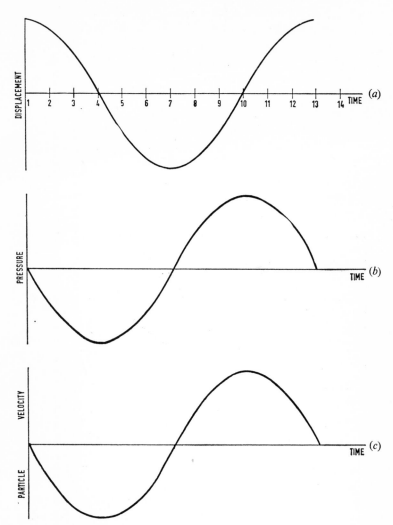

FIGURE 1.5

If we consider how one air particle is displaced we can find out how the pressure on the particle and its velocity vary.

(*a*) Shows the displacement at various times. For instance at 1 the particle is at maximum displacement. At 4 it is at its rest position, while at 7 it is again at maximum displacement but this time in the opposite direction to that at 1.

(*b*) The pressure at times 1, 7 and 13 is the steady atmospheric pressure value. At 4 the particle is the centre of a rarefaction while at 10 it is the centre of a compression.

(*c*) The particle can be likened to a pendulum, swinging first to one side and then to the other; its velocity varies with time as shown.

In Fig. 1.4*d* we showed how the pressure wave was set up in space. By utilising this information we can also indicate how the pressure on an individual particle varies during the displacement cycle. Tabulating the pressure conditions for various times of the displacement cycle we get:

Time 1 Normal
Time 3 Compression
Time 5 Normal
Time 7 Rarefaction
Time 9 Normal

Fig. 1.5*b* shows how the pressure varies.

Still considering an individual particle, an important point is how the velocity of the particle varies. Since velocity is the rate at which

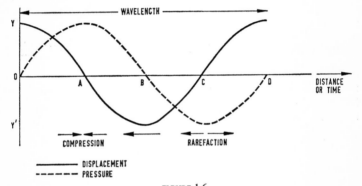

FIGURE 1.6

Showing how compressions and rarefactions alternate in the path of a sound wave. The wavelength is seen to be the distance between successive points which are at the same phase in the cycle.

the particle changes its position with respect to time, we can show the particle velocity directly on Fig. 1.5*c*. This particle velocity is, of course, superimposed on the random movement of the particles which occurs in any gas such as air. The particle velocity due to a sound wave is smaller than the velocities of this random movement.

Amplitude

Amplitude is the maximum value of a wave motion. For example, the maximum distance a particle moves from its rest position is the amplitude of displacement.

Wavelength

If we again consider the wave laid out in space, the distance between, say, two compressions, or between two rarefactions, is termed the *wavelength* (λ). And if we think of the wave passing a given point in space, and start counting just as a compression is passing this point, after a certain time has elapsed another compression will pass. This time is called the *periodic time* (T).

Frequency

The *frequency* of the wave is the number of complete changes, for example from compression to compression, which pass the given place in one second. It is measured in cycles per second (c/s) and given the symbol f. Frequency and periodic time are related as follows:

$$\text{Frequency} = \frac{1}{\text{Periodic Time}}$$

or

$$f = \frac{1}{T}$$

Velocity of Sound

The sound wave travels through the air with a certain velocity, and this velocity can be related to wavelength and periodic time as follows:

$$\text{Since Velocity} = \frac{\text{distance travelled}}{\text{time}}$$

$$v = \frac{\lambda}{T} = \lambda \times \frac{1}{T}.$$

But

$$\frac{1}{T} = f$$

$$\therefore \quad v = \lambda f.$$

(It should be noted that v is the velocity at which the wave travels through the air; it must not be confused with the *particle velocity*.)

For normal conditions the generally accepted value is 344 metres/sec or 1,130 ft/sec.

Among the factors which determine the velocity is the temperature. At 0° C the velocity is 331 metres per second or 1,086 feet per second. With a rise in temperature the velocity increases, the following formula relates them:

$$Vt = Vo\sqrt{1 + \frac{t}{273°}}$$

where Vt is the velocity at temperature t.
Vo is the velocity at 0° C.

When t is small compared with 273°, which it usually is, an approximation gives:

$$V = 33,100 + 60t.$$

Thus for each degree rise in temperature the velocity goes up by 60 cm per second. For comparison the velocities of sound waves in other media are quoted.

Oxygen   317 metres per sec.   Salt Water 1,504 metres per sec.
Hydrogen 1,270 metres per sec.   Steel   5,000 metres per sec.

It is interesting to calculate the wavelengths of two frequencies, one low the other high, which lie within the audible frequency range. At 50 c/s the wavelength is 22 ft 7 in.—at 15,000 c/s the wavelength is 0·9 in. This large range of wavelengths complicates the design of microphones and loudspeakers (see pp. 88 and 197).

Pitch

Up till now we have been discussing the physical properties of sound waves. Frequency, wavelength and velocity are the objective components of a sound; they can be measured. But the effect of a sound on a listener must be subjective, and here we will consider pitch.

Pitch is defined as that subjective quality of a note which enables one to place it on the musical scale. Pitch and frequency must obviously be closely related. The frequency of a sound depends on the rate of vibrations of the source. If the rate of vibrations is altered, the frequency will change and so will the pitch. Although pitch is mainly determined by frequency, there are circumstances where the pitch of a note can be altered by changing its intensity while keeping the frequency constant. We shall examine this dependence of pitch on intensity later in the book (see p. 40).

The pitch of a sound can also change if there is relative movement between the source and the listener. This effect is a very common one and most people will have observed it at one time or another. An express train sounding its whistle continuously as it enters a station is the example usually quoted, but there are others—a passing motor car for example.

The fundamental point is that when the distance between a source of sound and a listener is decreasing, the distance between two compressions—or between two rarefactions—is being reduced. This must mean that the effective wavelength is being decreased, and so the apparent pitch increases. Similarly, when the distance between the source and listener is increasing, the wavelength must be in-

20

creased and the apparent pitch drops. Thus, the pitch of a train whistle note drops just as it passes; the pitch as it recedes being lower than when it was approaching.

This change of pitch with change of distance between a source and a listener is called the Doppler effect.

Spherical and Plane Waves

So far we have considered how a sound wave travels through a small part of the air, but of course the wave must affect a lot more air than this. For example, we can begin by considering a source of sound which is radiating equally in all directions. This is known as a point source. If at a short distance from the source we think of how the sound energy is travelling, we can see that the wave has a 'wavefront' which is curved. In fact the total energy is being radiated over the surface area of a sphere. Hence the term 'spherical waves', the wavefront taking the form of a continually expanding sphere. In practice, a true point source is an impossibility. However, if a source has dimensions which are small compared with the wavelength, approximately spherical waves are radiated.

At very great distances from a point source, it can be seen that the curvature of the wavefront flattens out. The wavefront over a small area can be regarded as a plane at right angles to the direction in which the sound wave is travelling. This is referred to as a 'plane wave'.

Intensity

This is the rate at which energy is transferred through the medium, or the power passing through unit area—usually 1 sq. cm—of the wavefront. An easy way to look at this is to consider again our point source radiating in all directions, the power of the source being measured in watts. At a small distance from the source this power must be distributed over the surface area of a sphere. Now the surface area of a sphere is $4\pi r^2$ where r is the distance from the source or simply the radius of the sphere.

Therefore Intensity $= \dfrac{\text{Power of Source}}{4\pi r^2}$ watts/sq. cm.

Intensity and loudness are obviously related—more energy the louder the sound. Although loudness will be dealt with more fully in the next chapter it is interesting to note at this stage that the minimum intensity just to hear a 1,000 c/s note is 10^{-16} microwatts/sq. cm.

From the formula quoted above we can appreciate an important law of physics on how energy is distributed from a source. If we have a source of constant power then we can see that:

$$\text{Intensity} \propto \frac{1}{r^2}$$

Put into words—the intensity falls off inversely as the square of the distance. This is known as the Inverse Square Law. The source

FIGURE 1.7

A point source radiates equally in all directions. If *B* is twice as distant as *A*, the power is distributed over four times the area. Thus the intensity is inversely proportional to the square of the distance from the source.

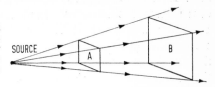

need not of course be a sound source—it could be a light source and this law indicates how the illumination would fall off as one moves away from the source. Fig. 1.7 shows pictorially how increasing the distance from the source increases the area over which the energy is distributed.

The intensity and pressure in a sound wave must be related and it can be shown that intensity is proportional to the square of the pressure or stated mathematically:

$$\text{Intensity} \propto \text{Pressure}^2$$

If we now substitute Pressure² for Intensity in the Inverse Square Law formula we get:

$$\text{Pressure}^2 \propto \frac{1}{r^2}$$

or
$$\text{Pressure} \propto \frac{1}{r}$$

If we draw a graph relating pressure and distance from the source (see Fig. 1.8) we find that, close to the source, a small increase in distance from *A* to *B* gives quite a large drop in pressure. If, however, we increase the distance by the same amount but this time start at *C* and move to *D*, the change in pressure is much less. This effect is of importance when using microphones, and we shall discuss it more fully in Chapter 6.

22

Intensity and Decibels

There is an enormous range of intensities in sound waves, something like 1,000 billion to 1. In most cases it is not the actual intensity value of a note that is important, but how it compares with other notes. This comparison has actually been defined by a law on hearing, the Weber–Fechner, which states that the aural effect of a change in intensity depends on the intensity preceding the change. A simple way of looking at this is to start with a note of a given intensity, say 10 units, and increase it to 100 units and then to 1,000 units. These two changes would be interpreted by the ear as equal changes in loudness since the ratio of $\frac{100}{10}$ is the same as the ratio of $\frac{1,000}{100}$.

Another way of stating this is to write the intensity values in powers of ten: 10^1, 10^2, 10^3 and it can be seen that equal changes in loudness are given by equal change in the *logarithm* of the intensity. The ear is said to have a logarithmic response, and a unit termed the BEL is

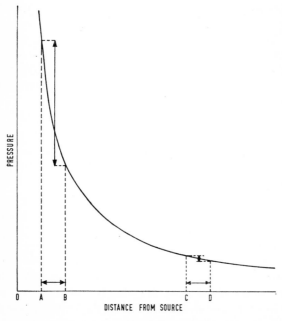

FIGURE 1.8

Fig. 1.7 showed how the intensity falls as the distance from the source is increased. Since intensity is proportional to the square of the pressure, the pressure will be inversely proportional to the distance. By plotting pressure against the distance we get the curve shown above. Note how, for a given change in distance, there is a much larger change in pressure at points close to the source.

23

used in acoustic measurements. If the initial intensity is I_1 and it is increased to a new value I_2 then:

$$\text{Ratio in BELS} = \log_{10} \frac{I_2}{I_1}$$

In practice the Bel is too large and a smaller unit the DECIBEL is used. The ratio in decibels is then:

$$10 \log_{10} \frac{I_2}{I_1}$$

Logarithms, are, of course, a very useful tool in calculation and the decibel is widely used in electronic engineering to compare two electrical powers. For example if the input power to an amplifier is P_{in} and the output power is P_{out} then the amplification in decibels is:

$$10 \log \frac{P_{out}}{P_{in}}.$$

Often in acoustics one deals with pressures rather than intensities and with two pressures the ratio in decibels is:

$$20 \log \frac{\text{Pressure}_2}{\text{Pressure}_1}.$$

This relationship is arrived at as follows:

$$\text{Decibels} = 10 \log \frac{I_2}{I_2}$$

$$\text{But Intensity} \propto \text{Pressure}^2$$

$$\text{Decibels} = 10 \log \frac{(\text{Pressure}_2)^2}{(\text{Pressure}_1)^2}$$

$$= 20 \log \frac{\text{Pressure}_2}{\text{Pressure}_1}$$

Travelling and Standing Waves

So far we have been considering the sound waves which travel freely outwards in space and meet no object which disturbs the even distribution of sound energy. But this is not what happens in practice. In a room, for example, the sound waves are reflected by the walls, ceiling, floor and the various objects in the room. When a wave is reflected back along its original path, it will interfere with the incident wave coming from the source. The interference pattern which is set up has a great effect on the distribution of sound energy. Of course

24

the intensity and distribution of the *reflected* sound depends on the size and shape of the reflecting surface or object, as we shall see later.

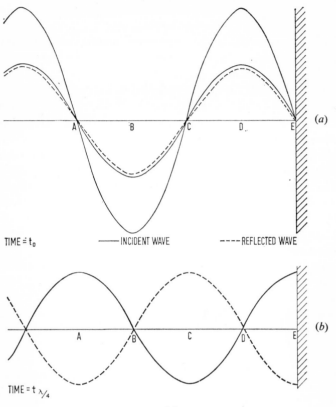

TIME = t_0 ———INCIDENT WAVE ----REFLECTED WAVE

TIME = $t_{\lambda/4}$

FIGURE 1.9

For a simple illustration of how standing or stationary waves are built up we can send a displacement wave down a rope which has one end fixed to a wall. This wave will be reflected back along the rope and, by examining the displacement of the rope at various points and at various times, we can see how the combination of the reflected and incident waves affect the rope.

(*a*) The first time chosen is when the two waves are in phase giving large displacements at *B* and *D*. This time we call t_0.

(*b*) If we make the time interval equal to that required for the two waves to travel a quarter of a wavelength, we get the two waves out of phase—note the waves travel in opposite directions. The resultant displacement is zero at all points of the rope. This time is $t_{\lambda/4}$.

A simple way to start looking at this problem is to consider the displacement wave pattern set up on a rope which has one end fixed rigidly to a wall. The free end is made to vibrate so that a wave of displacement travels down the rope to the wall. If we now make the

25

reasonable assumption that the wall is rigid, the energy in the rope must be completely reflected back along the rope. Now let us look at what happens to different parts of the rope at different times. Then we can build up a picture of what happens after sufficient time has elapsed for steady conditions to be set up.

Fig. 1.9a shows the conditions at time t_0. The dotted line represents the rest position of the rope, and is included for reference. The full line represents the wave coming from the driven end—the incident wave. The dashed line represents the wave coming from the wall—the reflected wave. At the instant of time we have chosen, the incident wave produces maximum displacement in the upward direction, quarter of a wavelength ($\lambda/4$) away from the wall. The reflected wave is in phase with the incident—their maxima occur at the same upwards and downwards position—and so the total displacement is as shown in the sketch by the broken line. It should be noted that, although we are assuming 100% reflection, the reflected wave is drawn slightly reduced in amplitude for clarity.

Now let a quarter of a cycle elapse—or put another way let both waves advance by a quarter of a wavelength. Looking first at the incident wave this means, for example, that the rest position at C moves forward to D or the maximum displacement at D moves forward to E. Fig. 1.9b shows these conditions, but the resultant conditions must obviously always produce no displacement at the wall, since the rope is rigidly fixed at that point. This means that the reflected wave must be returned in the opposite phase to the incident, so that the displacement at the wall is zero. And Fig. 1.9b shows how the reflected wave has moved forward (remember it travels in the opposite direction to the incident wave) so that it cancels at all points of the rope any displacement due to the incident. In fact, at this instant of time all parts of the rope are at the rest position.

Fig. 1.9c shows the conditions when the waves have advanced another quarter wavelength—a total of half wavelength ($\lambda/2$) since the start of our examination. We can see immediately that the conditions are similar to Fig. 1.9b; the two waves are in phase again. This time, however, the displacements are reversed; at D, for example, we have displacement in the downwards direction and at B in the upwards direction.

Fig. 1.9d shows the conditions after another quarter wavelength, a total of three-quarters of a wavelength ($\lambda/4$) since the start. And we again see that all parts of the rope must be at the rest position. The incident wave downwards displacement at the wall must be cancelled by the reflected wave and, since the latter has moved forward by $\lambda/4$, the cancellation takes place.

If we move forward another quarter of a wavelength, we shall have

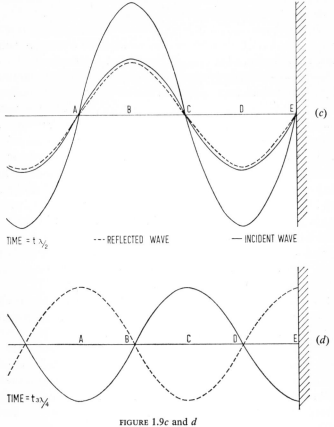

FIGURE 1.9c and d
(c) After anothert ime interval, at $t_{\lambda/2}$, the two waves are again in phase as at t_0 but with the displacement reversed.
(d) At $t_{3\lambda/4}$ the displacement is again zero all along the rope, since the two waves are again out of phase.

moved a total of one wavelength since we started, and the cycle will start over again. So the four sketches, taken at different times in one cycle, are all we need.

Over the whole cycle it can be seen that point B vibrates about the rest position of the rope. The same applies $\lambda/2$ away at D, but in the opposite direction—when B is above the rest position D is below, and vice versa. However points A, C and E *never* move. Point E is, of course, firmly fixed to the wall, but it is interesting to note that

27

at multiples of half-a-wavelength away, at points C and E, other parts of the rope never move.

One can see now why this is called a stationary or *standing wave*— it does not progress down the rope. Certain parts of the rope vibrate, while other parts do not move. In a *travelling wave*, all parts of the medium vibrate.

A point of no displacement, e.g. C, is called a *node*. A point of maximum displacement, e.g. D, is called an *anti-node*.

Suppose now that the rope is not fixed to a rigid wall, but to some object which can vibrate slightly. Then the reflected energy obviously cannot be equal to the incident, since some of the incident energy will be lost. The effect is to make the resultant wave amplitude at the anti-nodal points not twice the amplitude of the incident wave, as shown in Fig. 1.9, but some value which will depend on the amount of energy reflected. Similarly there will not be complete cancellation at the nodal points, as the two waves are no longer of the same amplitude. There will be a small resultant vibration, again depending on the amount of energy reflected. This will result in a series of maxima and minima.

If now we apply the same ideas to a sound wave, we will obtain some results which will be useful in later chapters on musical instruments, room acoustic and absorbent materials.

Let us start by again assuming a rigid wall which reflects all the incident energy. There are three 'constituents' to be considered— displacement, pressure and particle velocity.

Taking *displacement* first, we can see immediately that the standing wave pattern will be the same as the rope—a node at the wall, another $\lambda/2$ away and so on—an anti-node $\lambda/4$ away from the wall, another $3\lambda/4$ away and so on. Fig. 1.10a gives the pattern, and it should be noted that the diagram shows the variations in displacement for the various air particles in space. At $\lambda/4$ away from the wall, for example, we have maximum displacement—first to one side of rest position and then to the other. Fig. 1.10a is the usual method of showing a standing wave pattern.

Now consider *pressure*. If we take, for example, the point $\lambda/2$ away from the wall, i.e. at a node of displacement, the pressure at that point must be the centre of a compression, when the particles on either side move towards it, and be the centre of rarefaction when the particles move away. In other words the pressure at this point must be varying as shown in Fig. 1.10b and it must be an anti-nodal point. Similar reasoning—except of course that particles of air are only on one side—gives an anti-node of pressure at the wall. Be-

28

tween these two anti-nodes must be a node of pressure, at that point in space when the air particles have maximum displacement.

Lastly, we take *particle velocity*, and a moment's consideration

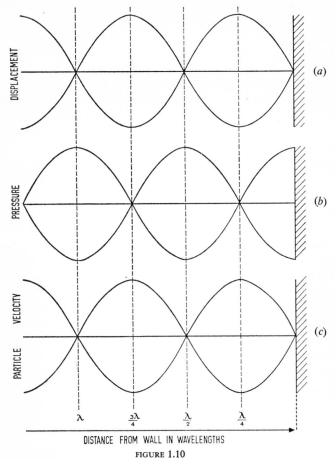

DISTANCE FROM WALL IN WAVELENGTHS

FIGURE 1.10

From the sketches in Fig. 1.9 we saw how the displacement standing wave was set up on a rope. Applying these ideas to our air particles and, assuming for ease of explanation that the reflected wave travels back along the original path, we get an air displacement standing wave pattern. Nodal points are at the wall, $\lambda/2$ and λ away from the wall. Anti-nodal points are at $\lambda/4$ and $3\lambda/4$ from the wall. From the displacement pattern (a) we can deduce the patterns for pressure and particle velocity (b) and (c).

will show that its standing wave pattern in space must be in phase with the displacement pattern. At positions of a node of displacement, the velocity must be zero since the particle does not move. Similarly, anti-nodes of displacement must also be positions of anti-nodes of particle velocity. Fig. 1.10c shows the particle velocity standing wave pattern.

Summarising the details of a sound standing wave system, one can say that at certain points in space the 'constituents'—displacements, pressure and particle velocity—vary with time, the anti-nodal points: at other points there is no variation with time—the nodal points.

As with the rope, of course, if the wall does absorb sound energy the standing wave pattern will be altered. The degree of alteration depends on how much energy is absorbed and hence how much energy is reflected.

CHAPTER 2

HOW WE HEAR

ALTHOUGH much work has been carried out on the hearing system, it would be true to say that even now it is not completely understood. However, a great deal of information is available that goes a long way towards explaining the mechanism of how we hear and the characteristics of the system—its frequency response, etc.

Physiological Details of the Ear

The basis of the system is that the variations of air pressure caused by the sound wave are converted into mechanical vibrations which in turn cause vibrations of a fluid. These fluid vibrations stimulate nerve endings which generate electrical pulses, and these are passed to the brain to produce the sensation of hearing.

This is the general principle of the mechanism. We can see that the physical parts of the system, or the anatomy, divide conveniently into three main parts:

1. Acoustical.
2. Mechanical.
3. Fluid/Electrical.

In studying the ear, it is convenient to divide it into three parts, referred to as the outer ear, middle ear and inner ear (see Fig. 2.1).

The outer ear starts with the external part of the ear which is readily visible, called the pinna. In human beings this has lost its function of helping to collect the sound waves, but with certain animals it is still used. Leading inside from the pinna is a canal about 25 mm long and 7 mm in diameter. This is closed at its inner end by the ear-drum, a thin membrane which vibrates when a sound wave strikes it.

The ear-drum is the first step in the mechanical system and marks the start of the inner ear. Attached to it is a bone called the 'hammer'. This is connected by ligaments to a second bone called the 'anvil'. The 'anvil' is attached to the third and last bone, called the 'stirrup'. These three bones are collectively termed the ossicles and they are extremely small. Their total weight is only about 63 milligrams and their size is roughly that of the head of a matchstick. The

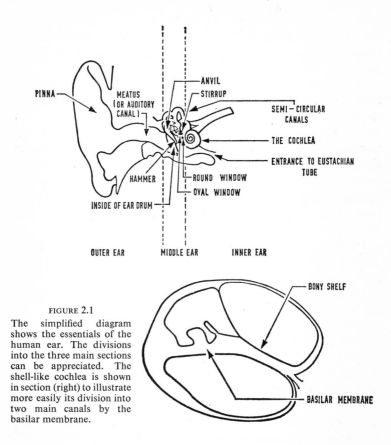

FIGURE 2.1

The simplified diagram shows the essentials of the human ear. The divisions into the three main sections can be appreciated. The shell-like cochlea is shown in section (right) to illustrate more easily its division into two main canals by the basilar membrane.

ossicles are housed in an air-filled cavity which is connected to the throat by the Eustachian tube. This acts as a pressure equalising device, maintaining the air pressure in the inner ear at the steady atmosphere value. This pressure will act on the inside of the eardrum, the outside will be affected by the sound wave pressure and the difference between these two pressures, due to the sound wave, will cause the ear-drum to vibrate.

The last of the ossicles, the stirrup, has its footplate fitted with ligaments into an opening called the oval window. This marks the junction between the middle and inner ears. The inner ear is a cavity window with bony walls filled with fluid. Just behind the oval window there is a space termed the vestibule leading to the end organ of hearing, the 'cochlea'.

The cochlea is shaped rather like a small spiral shell with $2\frac{3}{4}$ turns. The centre of this spiral has a perforated bone through which the auditory nerve passes leading to the brain. Before doing so, it joins with the nerve coming from the semi-circular canals, which are also situated in the inner ear. They have nothing to do with hearing, but provide us with our sense of balance.

The cochlea has three chambers, or canals, which are called the scala vestibule, the cochlea canal (sometimes the scala media) and the scala tympani. The scala vestibule is a continuation of the vestibule into which the oval window fits. Between the scala vestibule and the scala media there is a very thin, flexible membrane, and between the scala tympanium and the scala media there is the basilar membrane. Because the very thin membrane does not obstruct the passage of sound waves, one can imagine the cochlea to be essentially made up of two canals—the vestibule and tympanium—separated by the basilar membrane.

The basilar membrane does not quite separate the two canals. There is a small opening which joins them at the end remote from the oval window. The scala tympanium is, like the scala vestibule, connected into the middle ear, this time via a round membrane called, the 'round window'.

With the oval window being driven by the stirrup and the round window being able to vibrate and since the two canals are connected together, one can see that the fluid will be set in motion. This affects the basilar membrane, which has a complicated array of hair cells attached to it, and these are stimulated when the basilar membrane moves. The stimulation of a particular group of hair cells depends on the frequency of the sound wave, and this shows that the cochlea acts initially as a frequency analyser. When the intensity of a particular frequency is increased, not only does the associated point of the basilar membrane vibrate more but there is a spread of movement on both sides of this point as the intensity increases. The sensation of loudness seems to depend on the *number* of nerve pulses which are passed to the brain.

These cells produce the electrical potentials which control the signals sent along the nervous system to the auditory receptive centres of the brain.

Some dimensions will show how small are the cochlea and its constituent parts. The cochlea itself is about 5 mm from base to apex and 9 mm across the widest part. If straightened out, its length is about 35 mm. The area of the stirrup fitting into the oval window is about 3 mm², and that of the opening between the two canals is 0·25 mm². Its cross-sectional area varies, being widest at the vestibule end and narrowest at the apex. The basilar membrane has a width of about 0·25 mm and a length of approximately 32 mm, but this can vary slightly. There are several thousands of hair cells and these are laid out along the basilar membrane on a frequency basis, the high frequency ones being near the vestibule end while the low frequency ones are at the apex end.

The ear can detect sounds over a wide range of frequency and intensity. It can also detect very small changes in both these factors.

The foregoing paragraphs are a very brief and simplified account of the mechanism of the hearing system; now let us examine the characteristics of the system.

Aural Sensitivity

If one takes a tuning fork and sounds it, the waves it produces will be immediately audible. Gradually the loudness will get less and less until it cannot be heard. Yet, if one touches the fork lightly with the fingers, one finds that the fork is still vibrating. The fork must still be causing the air to vibrate, but since we cannot hear it there must be a minimum level of intensity below which we cannot hear. This intensity level, which divides sounds which we can hear and those which we cannot, is called the 'Threshold of Hearing' (see Fig. 2.2).

Suppose we have a source of sound and we can increase its intensity over a wide range. Starting at the threshold of hearing, increasing the intensity will increase the loudness. This process will continue until we find that the listening gets uncomfortable and eventually painful. There is thus a level which separates sounds heard properly by the hearing system and those which are felt or are uncomfortable. This level is called the 'Threshold of Feeling or Pain'. The two thresholds mark the limits of proper hearing. Outside them there is either no sensation of hearing or the hearing is painful.

With a subjective characteristic such as hearing, it is obvious that these two levels are not absolutely fixed. They vary from person to person. To get a representative figure, many people are tested and averaged curves are prepared to give the general picture.

When the threshold of hearing is examined, using pure tones, i.e. notes of a single frequency, it is found that the minimum intensity of sound required just to hear the note differs widely over the audible frequency range.

The ear is most sensitive in the region of 1,000 c/s to 5,000 c/s. For frequencies above this range, the sensitivity falls until, at some very high frequency, increase of the sound intensity does not produce

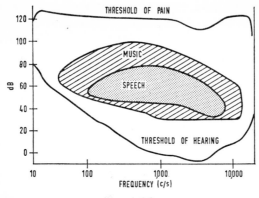

FIGURE 2.2

The large pressures required just to hear low and high frequencies compared to those necessary at middle frequencies can be appreciated from the curvature of the threshold of hearing. The threshold of pain is not so frequency dependent. Aural sensitivity curves such as these are obtained by using pure tones. In everyday life this full range is not exploited, and included in the diagram for comparison are the approximate frequency ranges for music and speech. The reference level is 0·0002 dyne/sq. cm.

any sensation of hearing. Below 1,000 c/s, increased intensity is required to hear the note and again we reach a low frequency which we cannot hear irrespective of its intensity.

The frequencies which mark the limits of audibility vary considerably, not only from person to person, but from age to age. As one gets older the hearing becomes impaired. The maximum range is usually quoted as approximately 20–20,000 c/s for adults; a range of about 30–17,000 c/s would be more representative.

Increasing age leads to a marked loss of sensitivity at the high frequencies. The low frequency sensitivity also falls, but the decrease is not so marked. Experiments show that the loss of high frequency sensitivity with age for men is greater than for women. This is balanced by the fact that women show greater loss at low frequencies.

The intensity required to produce pain does not show the same variations with frequency as does the threshold of hearing. The threshold of pain, although having some slight variations, is more uniform.

Since these two thresholds are not parallel lines, the intensity range over which the ear operates will vary with frequency. At the most sensitive frequencies the intensity range is wide, while towards the limits of the frequency range the intensity difference between the two thresholds is small. At its maximum, the intensity range is from about 10^{-16} to 10^{-4} watts/sq. cm, a range of 10^{-12} watts/sq. cm.

Recalling the section in Chapter 1 on decibels, we can see that this range of intensity can be stated in decibels as follows:

$$\text{Ratio in decibels} = 10 \log_{10} \text{Intensity Ratio}$$
$$= 10 \log_{10} \frac{10^{-16}}{10^{-4}}$$
$$= 10 \log 10^{-12}$$
$$= 10 \times 12$$
$$= \underline{120 \text{ dB}}$$

Loudness

Loudness is defined as the magnitude of the auditory sensation which a sound produces. We have seen that the threshold-of-hearing intensity varies over the frequency range, and this shows that the loudness of a sound does not solely depend on its intensity. If we took a note of 1,000 c/s at the threshold of hearing and increased its intensity, its loudness would increase. Say we increased the intensity from 10^{-16} watts/sq. cm to 10^{-13} watts/sq. cm, i.e. through a range of 30 dB, the loudness would obviously increase to a certain value. However, if we now changed the frequency of the note from 1,000 c/s to 100 c/s, we would find that it was just audible. This intensity level, 10^{-13} watts/sq. cm, is the threshold of hearing for the lower note.

From this we see that loudness depends on frequency as well as intensity. It is also evident that the decibel cannot be used in loudness measurement, unless modified in some way, since it is a ratio of two intensities. In assessing loudness, we must obviously use subjective techniques, the listener judging the effect of changing the intensity or the frequency. This leads to a unit of loudness level called the 'Phon'.

36

The Phon

In using the phon, the technique is to compare the loudness of a reference note with the loudness of the given note. The intensity of the reference note can be adjusted until it sounds as loud as the given note. The amount by which the reference note has been moved away, from its threshold-of-hearing intensity value, gives the loudness level of the given note in phons (see Fig. 2.3).

It is necessary to specify not only the frequency of the reference note, but also its threshold-of-hearing intensity value. The reference

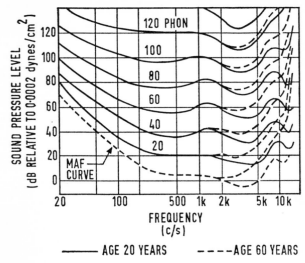

FIGURE 2.3

Equal loudness contours derived from work carried out by Robinson and Dadson. This work has produced results which differ to some extent from that done by Fletcher and Munson in 1933 on which Fig. 2.2 is based. One difference is that no O phon curve is included as the threshold of hearing has been found to be about 4 dB above 0·0002 dynes/sq. cm. The lowest curve refers to the minimum audible field. (By courtesy of Dawe Instruments Limited.)

note used is a pure tone of 1,000 c/s and the reference intensity is 10^{-16} watts/sq. cm, which corresponds to a pressure of 0·0002 dynes/sq. cm.

With normal listening conditions, an observer is readily able to judge when the loudness of the 1,000 c/s note is the same as the given note. To obtain this parity, the 1,000 c/s note intensity will have been raised by a certain number of decibels above the reference intensity value. The number of decibels gives the equivalent loudness—or loudness level—of the given note in phons.

Another way of putting this is to say that at 1,000 c/s, where there are 120 dB between the thresholds, there are 120 phons. The phon and decibel are taken to be equal at the reference frequency.

Using this technique, it is possible to prepare loudness level curves which join points at different frequencies producing equal

TABLE 2.1. *Relative Levels of Typical Sounds*

Noise	dB	Relative energy	Sound pressure dynes/cm²	Typical examples
				Threshold of pain
	– 120 –	–1,000,000,000,000–	– 200 –	
Deafening	– 110 –	– 100,000,000,000 –		Thunder Gunfire Pneumatic drill Steam whistle Large machine shop
	– 100 –	– 10,000,000,000 –	– 20 –	
Very loud	– 90 –	– 1,000,000,000 –		Underground railway Busy street Noisy factory Inside aeroplane Loud public address system
	– 80 –	– 100,000,000 –	– 2 –	
Loud	– 70 –	10,000,000		Noisy office Suburban train Typewriters Radio set—full volume Average factory
	– 60 –	– 1,000,000 –	– 0·2 –	
Moderate	– 50 –	– 100,000 –		Large shop Average office Quiet motor car Quiet office Average house
	– 40 –	– 10,000 –	– 0·02 –	
Faint	– 30 –	– 1,000 –		Public library Country road Quiet conversation Rustle of paper Whisper
	– 20 –	– 100 –	– 0·002 –	
Very faint	– 10 –	– 10 –		Quiet church Still night in the country Sound-proof room Threshold of hearing
	– 0 –	– 1 –	– 0·0002 –	

loudness. The 'O' phon line is the threshold of hearing, since here all frequencies are just audible. Other phon values are plotted to cover the range of hearing. The phon curves are often referred to as 'equal loudness contours', the method of marking being similar to the way in which contours of equal height are shown on a map.

These phon curves show that, if the intensity level is low, the range of frequencies which is heard is limited. The curves for the low values of loudness are just as curved as the O phon line, it is not until about 80 phons that the line becomes fairly straight. Adjustment of intensity will thus affect the relative balance between the various parts of the audible frequency range. For example, a low setting of the volume control on a record player will produce a different loudness pattern for the low and high frequency response than for a high setting.

The phon curves are also used in the design of some noise meters. Measurement of noise is often conveniently done by using 'subjective meters'. These have filters in their amplifier chain which give the meter an overall response similar to the pure-tone loudness contours. These sound level meters usually have a choice of three responses, depending on the strength of the noise. One corresponds to the 40 phons contour, the second to 70 phons and the third to 80 phons, which is used for all sounds above this loudness level. The meter is calibrated in decibels (see Fig. 2.4).

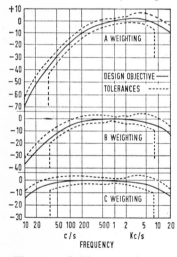

FIGURE 2.4
The curves show the response of weighting networks used in the measurement of noise. (By courtesy of Dawe Instruments Limited.)

The use of this type of meter does give readings which show the 'nuisance' value of noise. The meter is said to be 'weighted', just as the ear response is 'weighted' against the low and high frequencies.

Another example of the use of filters to simulate ear response is met in tele-communications. Measurement of the noise level on lines is done by using, at the receiving end, a 'weighting' network which again ensures that the decibel value of the measured noise indicates its nuisance value.

39

Pitch

Pitch was defined in Chapter 1 as being the subjective quality of a note which enables one to place it on the musical scale. Although it is mainly determined by the frequency of vibrations of the sounding body, it also depends on the intensity of the note.

With pure tones at low levels of intensity, the pitch depends only on frequency; that is the pitch of a note remains the same although the intensity is altered. However, with tones which are 40 dB above the threshold of hearing it is found for low frequency notes that there is a change of pitch if the intensity is altered. Increasing the intensity lowers the apparent pitch. If the frequency is variable, a listener can raise the frequency to restore the pitch when the intensity is increased. This effect is most marked around 200 c/s. With high frequency notes, the change goes the other way; increase of intensity raises the pitch of the note.

This change of pitch with intensity is confined to the low and high frequencies. With the middle frequencies, in the range 750 c/s to 6,000 c/s approximately, there is little or no change of pitch.

With notes combining several frequencies, from a musical instrument, for example, the appreciation of pitch is somewhat different. It is possible to remove or increase the lowest frequency without the pitch of the whole note being altered.

Masking

If, when we are listening to one note another note is sounded, the second note may not be heard although its frequency and intensity are inside the range of hearing. In other words, when the ear is being affected by one sound, its sensitivity to other sounds is decreased. This is called *masking* and the effect is easily noticeable.

The actual masking effect depends on whether pure tones or noise are acting as the masking tone. Experiments have provided data which show how much the intensity of a sound must be raised to be just audible when a masking tone or noise is present.

Directional Hearing

The question of how we can locate the direction of the source of a sound has exercised the minds of investigators for many years. The subject has received increased attention with the development of stereophonic recording and reproduction.

This faculty of directional hearing is a complex one depending on many factors, but a start can be made by noting that the acoustic impression of our environment is conveyed to our brain by two channels which, of course, begin with the ears. Since they are spaced apart, the two ears are affected slightly differently. This difference provides acoustic clues which can be interpreted by the brain, using its built-in store of acoustical experience, and allow it to ascribe a direction to the source.

How the brain interprets the clues is beyond the scope of this book, but research work done recently in Great Britain has produced much new knowledge of how the brain tests the clues against past experience, probabilities and habits. The attention of the interested reader is drawn to the references where several important papers on the subject are listed (page 216).

Let us see how some of the acoustic clues can be produced. An early attractive theory was that the signals at the two ears differed in *intensity*, the difference being produced by the shadowing section of the head. The listener locates the source as being in the direction of the ear receiving the greater intensity. An extension of this idea was that, where the person was unsure of the source's location, he turned his head until the sound and the ears were in the same line. This would then make the intensity difference a maximum and hence allow accurate location.

However, an object can affect the distribution of energy in a sound field only if its size is comparable to or greater than the wavelength of the sound. Regarding the head as an object, we see that at low frequencies, where the wavelength is large, the head will not cast sufficient shadow to cause any appreciable intensity difference between the signals at the two ears. At high frequencies, however, the head is able to cast a shadow and so provide the intensity difference. The approximate dividing line is 1,000 c/s.

It was Lord Rayleigh who pointed out that the intensity theory was, of necessity, only of limited use in explaining directional hearing. Although it might explain the effect at high frequencies, he saw that it could not be applied to low frequencies.

Lord Rayleigh's next step was to examine the possibility that *phase difference* provided the brain with the necessary directional clues. His experiments led him to discover that there is sufficient phase difference at low frequencies to cause the signals to differ by an appreciable amount. He then concluded that both intensity and phase differences are utilised, the particular system depending on whether the frequency is low or high.

Later working during the 1930's showed, however, that the effect of a phase difference could be regarded in the same way as that produced by a *time-of-arrival difference* and that the latter was more likely to be directly noticed by the listener. Further work showed that this time difference would also be effective at high frequencies and that the intensity difference theory was suspect because the differences actually experienced seemed insufficient to give the known accuracy of location.

More recent research in Great Britain has tended to confirm the theory that it is time-of-arrival differences which enable us to locate the direction of a source. It has also been able to confirm that the ability to locate a source with accuracy depends on the nature of the source. If the source is producing pure tones, the location is uncertain, but if the source produces complex waves more precise location is possible. Sources producing a very wide range of frequencies, such as noises or clicks, can be located very accurately.

So it would seem that our directional hearing ability depends to a great extent on time-of-arrival differences. However, it must be remembered that this is not the only factor; there are probably others of equal importance. As with hearing itself, although a great deal is known, further work is necessary before we fully understand how the combination of ears and brain enables us to locate with fair accuracy the position of a sound source.

CHAPTER 3

SOURCES OF SOUND

WE have seen that any vibrating body is a source of sound, providing the frequency and intensity of the sound waves it produces are inside the range of hearing.

Most sound sources operate in a complicated way. Take a violin or a piano, or someone speaking, for example, or the cone of a loudspeaker which has to vibrate in such a complicated manner when emitting sound waves reproducing the intricacies of a full orchestra. These very complex systems defy a full analysis, but we can at least examine the basic laws which govern individual properties of sounds.

Tone Quality

Pitch and intensity have already been discussed in Chapter 1, but this is a suitable point at which to explain what is meant by tone quality. Much of the pleasure in listening to various musical instruments is due to the fact that each type of instrument has its own 'quality' or 'colour' of tone. The word *timbre* is often used for this characteristic by which we can distinguish between the various instruments.

If we take a tuning fork and examine its output on an oscilloscope, we find that it has a smooth sinusoidal waveform. Suppose we now take a clarinet and sound the same note as the fork. Not only will it sound very different, but we shall find that the waveform is quite different too. Instead of the smooth shape produced by the fork, the wave shape is such that, although the cycles repeat in the normal way, the shape of each cycle is irregular. The sound of the fork is said to be 'pure' while that of the clarinet is 'complex'.

The note of the fork consists of only one frequency whereas the clarinet sound is a mixture of several frequencies. This difference

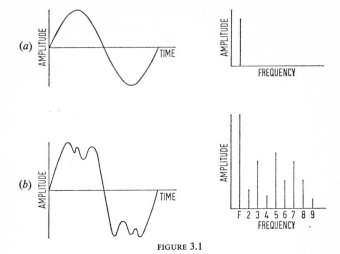

FIGURE 3.1

(a) The waveform of a pure note, e.g. from a tuning fork. The spectram diagram (right) shows the single line at the frequency of the note.
(b) The waveform of a clarinet shows a much more complicated wave and the spectram diagram gives the amplitudes of the fundamental and harmonies.

could be demonstrated by means of a spectrum analyser, an instrument which separates the frequencies present in a sound and can display them as spaced vertical lines. The height of each vertical line gives the amplitude of that particular frequency (see Fig. 3.1).

The fork will show only one line, at the frequency marked on the fork, but the clarinet produces several frequencies simultaneously and there will be a series of lines. The lowest frequency of the series will be the same as the fork; this is called the *fundamental*. The other frequencies are the overtones of the fundamental. The spectrum analyser would show that all the overtones do not have the same amplitude, some being stronger than others. It is the presence of overtones and their relative strengths which give the clarinet its characteristic tone.

Harmonics

Most musical instruments produce complex waves, each type varying in its overtone content and each thus having its own particular timbre.

44

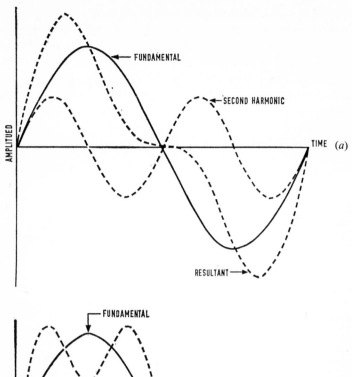

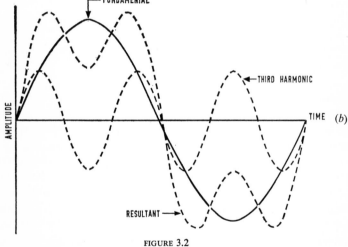

FIGURE 3.2

(a) The addition of a fundamental and its second harmonic gives a 'complex' wave. The exact shape of this resultant depends on the relative amplitude and phase of the fundamental and second harmonic. With complex sound waves the ear ignores changes of phase, so that even though the phase relation of the two components alters and the wave shape changes there is no difference in the tone quality. (b) The complex wave produced by adding a third harmonic to its fundamental has quite a different shape from that shown in (a).

The overtones in some cases are exact multiples of the fundamental and are then termed *harmonics*. String, wood-wind and brass instruments all produce a series of harmonics in this way. In some other instruments the overtones are not exact harmonics of the fundamental, for example the xylophone and tubular bells (see Fig. 3.2).

Another feature of the complex sounds from many instruments is that the time duration of the overtones varies. At the instant that the note is produced, many overtones are generated but the higher ones tend to be weak and die out quickly. The lower frequency overtones are generally much stronger and persist for a longer time. The fact that the higher overtones persist for only a short time produces a tone quality for the initial part of the sound which differs from that of the remaining or continuing part. 'Attack' is a word often used to describe this effect.

An interesting way of showing how important the attack and continuing tones are to the overall quality of an instrument occurs when producers of so-called 'electronic music' cut up tape recordings to obtain novel sound effects. If a piano is recorded on tape and the attack or strike tone is cut out, the resultant note is quite different from the original. A variation on this is to reverse the order by playing the continuing tone first. These simple transpositions offer limitless possibilities for producing new sounds.

Since the higher overtones making up the attack tone pass quickly, they are said to be transient in nature or simply transients. In acoustics there are many examples of transients occurring, rapid changes of amplitude with time, and since their presence is an essential for the musical timbre, any recording and reproducing system must be able to handle them with fidelity. A difficulty arises in the electro-mechanical transducers—microphones and loudspeakers—since the mechanical system may possess inertia. This can cause a 'hang-over', a continuation of vibration after the transient has passed. Modern microphones have practically no transient distortion but loudspeakers, with their larger moving systems, can introduce considerable distortion due to this cause.

The above brief introduction to tone quality or timbre will suffice at this stage to help in understanding the production of sounds by various vibrating methods. When discussing the instruments in detail, we shall go further into how the player can control the timbre.

Fundamental Features of Sound Sources

It should perhaps be pointed out that the sound sources under discussion here are those which produce pleasant tones—musical

46

instruments, the human voice, etc. Certain features are common to all such sound sources. First there is a method of excitation, then a resonator and, lastly, in many cases, a radiator. The excitation—fingers, air stream, sticks—set the resonant body into vibration and, since some resonators do not produce a powerful sound wave, they are coupled to a body which will vibrate in sympathy, move a lot more air and hence amplify the weak vibrations of the resonator.

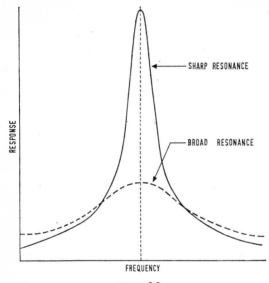

FIGURE 3.3

The degree to which a resonator responds depends on whether it has a sharp or broad resonance. A sharp resonance gives a large response at the resonant frequency with a rapid fall in response as the frequency moves away from resonance. Broad resonance gives a much reduced response at resonance and considerable response at frequencies off response. Where the exciting system has a wide range of frequencies the resonator should have a broad resonance to give general reinforcement.

The vibrating systems which are used as resonators can be listed as follows:

   (i) Air columns.
  (ii) Strings.
 (iii) Bars.
 (iv) Stretched membranes.
  (v) Circular plates.
 (vi) Vocal chords.

Air Columns

If a cylindrical tube is closed at one end and a means of excitation, a tuning fork for example, brought near the open end, the air in the tube will resonate if its length bears a certain relationship to the wavelength of the note sounded by the fork.

When the fork vibrates it causes compressions and rarefactions to be sent into the tube. Since one end is closed, these must be reflected back along the tube and the result of the two wave motions moving in opposite directions is to create a standing wave system.

In Chapter 1 we saw that, by knowing the conditions at the reflecting surface, we were able to get a picture of how the air was behaving at various points in space. If there was 100% reflection, there would be a condition of no displacement (a node) at the surface, but a quarter of a wavelength away the displacement would be a maximum (an anti-node).

Closed Pipe

If we now consider the pipe closed at one end, then obviously there must be a node at the closed end. At the open end the air is not constrained in any way and must be free to vibrate; hence this must be the position of an anti-node of displacement. These are conditions which must be met if the air in the pipe is going to resonate. One cannot, for example, imagine the air at the open end standing still; there must always be an anti-node of displacement at this point.

The distance between a node and the nearest anti-node is quarter-of-a-wavelength, so that this must be the length of the pipe when it is resonating (see Fig. 3.4).

i.e.
$$\text{length} = \frac{\text{wavelength}}{4}$$

$$l = \frac{\lambda}{4}.$$

Rearranging gives
$$\lambda = 4l.$$

Recalling that in any wave motion

$$\text{Velocity} = \text{Frequency} \times \text{Wavelength}$$

i.e.
$$c = f \times \lambda.$$

Substituting $4l$ for λ we get

$$c = f \times 4l$$

or
$$f = \frac{c}{4l} \text{ c.p.s.}$$

We see from this that the longer the pipe the lower the note—a not unexpected result.

The lowest note the pipe produces is termed the fundamental. It can simultaneously sound overtones which are harmonics of the fundamental, provided they meet the required conditions of the

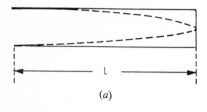

(a)

FIGURE 3.4
Vibration of air in a pipe closed at one end.
(a) This shows the displacement pattern in a pipe of length '*l*' when resonating at its fundamental frequency.

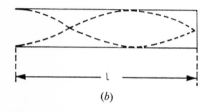

(b)

(b) The third harmonic will be sounded since the displacement pattern it produces meets the standing wave requirements. All the odd harmonics produce the required anti-node at the open end.

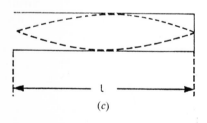

(c)

(c) To show that even harmonics are not possible, the pattern for the second harmonic is drawn and it can be seen that it would require a node of displacement at the open end, which is impossible.

standing wave system in the pipe. And straight away we see that the even harmonies cannot be sounded by a closed pipe, since they would require exactly the same conditions at each end of the pipe— an impossible condition. The second harmonic, for example, would require a node at the closed end, an anti-node in the middle and an anti-node at the open end. But the last condition cannot occur and so the second harmonic will not be sounded, neither will the fourth, sixth, etc. The third harmonic will, however, be supported since it

D   49 will provide an anti-node of displacement at the open end. In fact, all the odd harmonics—fifth, seventh, etc.—meet the conditions.

We thus see that a pipe closed at one end resonates when its length is a quarter-of-wavelength and that it supports only the odd harmonics.

Open Pipe

A pipe open at both ends will also resonate if it is excited by a vibrating body. Again a standing wave system is set up in the air, due to reflection from the open ends. These reflections occur because there is a change in conditions for the air at the open ends.

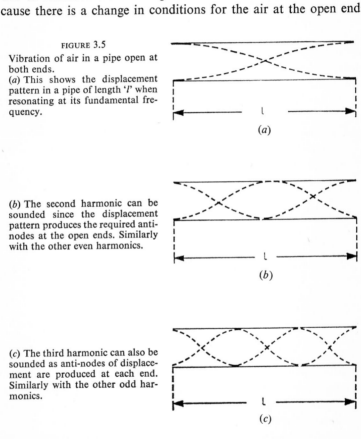

FIGURE 3.5

Vibration of air in a pipe open at both ends.
(a) This shows the displacement pattern in a pipe of length '*l*' when resonating at its fundamental frequency.

(a)

(b) The second harmonic can be sounded since the displacement pattern produces the required anti-nodes at the open ends. Similarly with the other even harmonics.

(b)

(c) The third harmonic can also be sounded as anti-nodes of displacement are produced at each end. Similarly with the other odd harmonics.

(c)

Inside, the air is constrained by the walls of the pipe; outside, the air is under no restraint other than that due to atmosphere pressure. At the boundary between the two conditions, reflection takes place.

Since the conditions are the same at both ends and the ends are

50

open, there must be anti-nodes of displacement at each end for the fundamental. The minimum distance between two similar conditions in a standing wave system is half-a-wavelength. The pipe will, therefore, sound the fundamental when its length is half-a-wavelength.

i.e.   when $l = \dfrac{\lambda}{2}$

Rearranging $\lambda = 2l$

Substituting for λ

$$c = f \times 2l$$

or $$f = \frac{c}{2l} \text{ c.p.s.}$$

We can see that if we have two pipes of the same length, one pipe closed at one end and the other open at both ends, the open pipe will sound an octave above the closed pipe. The open pipe supports all the harmonic series, both odds and evens, since they all meet the necessary conditions by providing anti-nodes of displacement at each end (see Fig. 3.5).

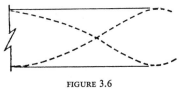

FIGURE 3.6

For simplicity in the previous figures it was assumed that the displacement node coincided exactly with the end of the pipe. In practice this is not so; the anti-node is always slightly outside the pipe as illustrated in the sketch. This means that the effective length of the pipe is longer than its physical length.
To allow for this an 'end-correction' is applied. The amount of end-correction depends on the radius of the pipe and the type of pipe. For a cylindrical pipe it is $0 \cdot 6R$ where R is the radius. For other shapes the correction depends on the degree of openness.

When the air in a pipe is vibrating, it continues to act as a column slightly beyond the ends of the pipe. The effective length of the column is thus larger than the physical length of the pipe. This error is taken into account by adding a length known as the 'end correction'. The amount of end correction depends on the radius of the pipe and also on the shape of the end, if other than a simple cylindrical pipe is used.

Variation of Pitch with Temperature
In the relationship between the pitch of the note and the length of the pipe we have assumed that 'c' the velocity of sound, is constant.

However, we saw in Chapter 1 that the velocity depends on temperature, rising when the temperature rises. For a given length, therefore, the pitch of the note will rise if the temperature rises. This effect can be quite noticeable and is obviously of importance to all instruments using an air column as the vibrating resonant system.

Strings

Normally we are considering here strings of flexible material stretched between rigid supports. When the string is made to vibrate transversely we have a resonant system widely used in musical instruments. As with the air column, a standing wave system is set up due to reflection from the fixed ends.

Since the ends are fixed, there must be nodes of displacement at each end. The simplest pattern of vibration therefore occurs when there is an anti-node in the middle. This must occur when the length of the string is half-a-wavelength long (see Fig. 3.7).

i.e. $$l = \frac{\lambda}{2}$$

Rearranging $$\lambda = 2l$$

Substituting for λ

$$c = f \times 2l$$

or $$f = \frac{c}{2l} \text{ c.p.s.}$$

This is the fundamental; the string will also support a series of harmonics. Since the conditions are the same at each end any harmonic will meet the requirements and the entire series of odd and even harmonics can be sounded. The actual harmonic content of the complex note produced by a string vibrating, i.e. the presence of particular harmonics and their relative amplitudes, depends on how the string is made to vibrate and where the exciting force is applied. We shall go further into this in the section on musical instruments in Chapter 4.

The velocity 'c' used in the above formula is the velocity of the wave along the string, and any variation in this will alter the pitch of the note—as happened with the air column. There we saw that the velocity depended on the temperature of the air. With the string, the velocity depends on the tension and the diameter of the wire. If we are comparing strings of different materials, the 'mass per unit length' is used instead of the diameter as it is not simply the diameter which is important, but the mass of the string.

We can see how the velocity depends on these factors by looking at a stringed instrument such as a violin. The pitch of a note increases when the tension is increased by screwing the peg, and it decreases when we vibrate the open strings in turn starting with the

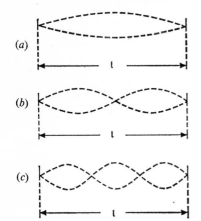

(a)

(b)

(c)

FIGURE 3.7
Transverse vibrations of a string.
(a) With the string clamped at both ends the fundamental produces a displacement pattern as shown.

(b) and (c) The entire harmonic series —even and odd—can be supported since they fit the requirements of the displacement standing wave.

thinnest one. The velocity (c), tension (T) and mass per unit length (M) are, in fact related as follows:

$$c = \sqrt{\frac{T}{M}}$$

Substituting for c, we get:

$$f = \frac{1}{2l}\sqrt{\frac{T}{M}} \text{ c.p.s.}$$

Summarising, therefore, we can see from this result that the pitch of a transversely vibrating string depends on:

1. *Length:* If the length increases the pitch decreases.
2. *Tension:* If the tension increases the pitch increases.
3. *Mass per unit length:* If the mass increases the pitch decreases.

Bars

The vibration of bars or rods can take place in three different ways— transversely, longitudinally and by torsion. The most interesting, since it is the mode of vibration used in some musical instruments, is the transverse mode. Reeds are examples of transverse vibrating bars; a tuning fork is another and the xylophone makes use of bars which are 'free', that is not fixed or restrained in any way.

53

Clamped Bar

We mean here a bar clamped at one end. The action is complicated and does not respond to simple analysis as in the case of the air column or strings. However, we know that the fundamental is sounded when the free end is an anti-node of displacement. The pitch

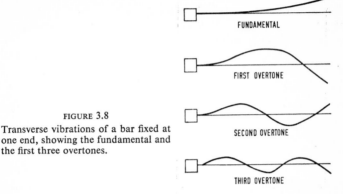

FUNDAMENTAL

FIRST OVERTONE

FIGURE 3.8

Transverse vibrations of a bar fixed at one end, showing the fundamental and the first three overtones.

SECOND OVERTONE

THIRD OVERTONE

is controlled by the length, the density and elasticity of the bar material and the radius of gyration. The radius of gyration depends on the type of bar, whether it has a rectangular or circular cross-section, or if it is hollow or solid.

The bar produces overtones where other parts of the bar, as well as the free end, have considerable displacement. However, unlike the overtones of a vibrating string, the overtones of the bar are inharmonic and are related to the fundamental, f, as follows:

$$1\text{st Overtone} = 6\cdot27f$$
$$2\text{nd Overtone} = 17\cdot55f$$
$$3\text{rd Overtone} = 34\cdot4f$$

Because these overtones are of much higher frequency than the fundamental, they contain little energy and are transitory.

The tuning fork is a common example of a transversely vibrating clamped bar and it is possible to demonstrate with a fork how fleeting are the overtones it produces. If it is struck at an anti-nodal point for an overtone, the overtone can sometimes be heard immediately the fork has been struck. But this soon dies away to leave the pure note of the fundamental.

Free Bar

Here we mean a bar unclamped in any way, supported at two points. As with the clamped bar, the frequency of the fundamental depends on the length, density, etc. Again overtones are produced, but their relationship with the fundamental is quite different as is shown below:

$$1\text{st Overtone} = 2\cdot76f$$
$$2\text{nd Overtone} = 5\cdot4f$$
$$3\text{rd Overtone} = 8\cdot9f$$

These are again inharmonic, but are much closer to the fundamental and contain more energy. They are therefore not so transitory as in the case of the clamped bar.

Stretched Membranes

The detailed study of stretched membranes is very complicated, but they are the vibrating structures in an important class of instrument —the drums. The diaphragm on a condenser microphone, for example, is also a stretched membrane. The membrane is flexible and stretched in all directions by a heavy surround.

The vibrations of a membrane are fundamentally different from, for example, the air column we have already discussed. There the bore, or diameter, of the column was much smaller than its length and the vibrations were longitudinal along the pipe. The air column system is comparatively simple since the air only vibrates along one direction or, more technically, it is a one dimensional system. A membrane, however, is not constrained in such a manner. It is a

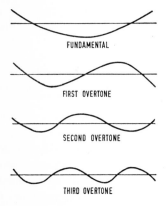

FUNDAMENTAL

FIRST OVERTONE

SECOND OVERTONE

FIGURE 3.9
The transverse vibrations of a bar free at both ends, showing the fundamental and the first three overtones.

THIRD OVERTONE

two dimensional system and so it can vibrate in more than one direction. This produces patterns of vibration which contain nodal (no movement) circles and nodal diameter (see Fig. 3.10).

The fundamental note is sounded when the membrane is producing an anti-node at the centre and, of course, the first nodal circle occurs at the edge where the membrane is rigidly clamped.

The pitch of this fundamental is determined by the radius, tension and mass of the membrane. Obviously, as either the radius or the mass increases, the pitch will fall; if the tension goes up, the pitch will rise. The use of the tension to control the pitch can be seen on

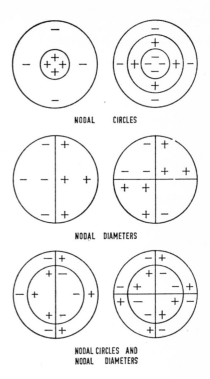

NODAL   CIRCLES

NODAL   DIAMETERS

NODAL CIRCLES AND
NODAL   DIAMETERS

FIGURE 3.10

The complicated modes of vibrations of membranes and plates are indicated by the sketches in which the differing directions of displacement are represented by the positive and negative signs.

the tympani when the player adjusts the tensioning screws on the surround.

The overtones of the fundamental are not harmonic. The first one relating to nodal circles will be the one which has two circles, and this has a frequency 2·3 times the fundamental. The first one having a nodal diameter is at 1·59 times the fundamental. Other overtones are produced by nodal circles and nodal diameters acting together.

When used as a microphone diaphragm, of course, it is essential that the membrane does not break up into nodal diameters or circles. Otherwise the diaphragm will not be moving forward and back as a whole. Should the direction of displacement be different for different parts of the diaphragm, the average displacement will be reduced, so lowering the output of the microphone.

56

This difficulty is overcome by making the fundamental frequency of resonance higher than the highest audio frequency to be reproduced. The diaphragm displacement then follows the shape of a parabola, all moving parts having moved in the same direction.

Circular Plates

These are somewhat similar to stretched membranes except for one important difference. We saw that a membrane is flexible and under tension. This tension provides what is called the restoring force, the force which makes the membrane return to its rest position when it is displaced. A string is another example where tension must be applied to get the flexible material to vibrate.

However, with a plate no tension is required as the plate material itself has the necessary stiffness to provide the restoring force. As with membranes and strings, an analogy is possible between plates and bars.

Like the membrane, the plate can vibrate in complicated patterns giving rise to nodal circles and nodal diameters. The fundamental frequency basically depends on the thickness of the plate, its radius and the material; but it also depends on how the plate is supported. The plate can be clamped at its periphery, supported at the edges or the centre, or be quite free. Overtones are produced which are not harmonic, their relation to the fundamental again depending on how the plate is supported.

The gong and cymbal are examples of instruments using plates. Another common example is the diaphragm plate used in telephone receivers.

As with the membrane type of diaphragm, it would be advantageous if the thin plate could be driven so that it did not break up into nodal diameters or circles. This, we saw, could be avoided by making the fundamental frequency high in relation to the sound frequencies. With thin plates, however, it is difficult to get a light diaphragm which will be sensitive and yet be stiff enough to give a high fundamental frequency. At one time, thin plate diaphragms were used in condenser microphones, but for this reason they have been almost superseded by membranes.

In telephone receivers, the frequency range required is much more restricted than that required from a condenser microphone and the non-uniform response of the thin plate diaphragm is unimportant in view of its simplicity and ruggedness.

Vocal Cords

The vocal cords are stretched across the top of the larynx which closes the trachea or windpipe, leading from the lungs. They are not cords exactly but are folds or tissues which vibrate and control the size and shape of the larynx opening.

When one is simply breathing, the cords are widely separated at one end and they form a triangular opening. To produce sounds the cords are brought closer and the flow of air from the lungs up to the trachea causes them to vibrate. The effect of this is to superimpose a fluctuation on the otherwise steady air stream flowing from the lungs into the throat. This is the start of the variations required to produce sound waves. The fluctuating air stream passes into the cavities of the throat, mouth and nose. These cavities resonate and thereby accentuate certain frequencies.

This is the action used when producing certain sounds; a general term for them is the 'voiced' sounds. Pure vowels—'a' for example—are sounded in this way.

With other sounds, however, the vocal cords have very little to contribute. Instead the air is made to pass through narrow openings and past sharp edges. This causes the air stream to have a fluctuating character and it is the air vibrations which are the source of sound. 'S' is an example of this second group which are called the 'unvoiced' sounds.

Resonators, Sounding Boards, etc.

With many sources of sound—a vibrating string for example—there is not a great deal of air movement due to the vibrator alone and the sound produced is weak. This can be improved by coupling to the primary source of sound a body which will resonate in sympathy and hence, by causing more air vibrations, produce an increased sound output.

Taking two instruments using strings as the vibrator, the violin and the piano, we can see in a simplified manner how the resonators behave.

With the violin, the strings are stretched across the bridge. When they vibrate they cause the air inside the body and the body itself to vibrate. With the piano, the strings are stretched by a steel frame across two bridges, one being on the frame itself, the other being on a soundboard. When a string is struck by the hammer, the resultant vibrations are coupled to the soundboard which in turn vibrates.

There are many other examples which illustrate the basic idea of sound re-inforcement by a resonant body. We saw that, with the voice, the cavities in the mouth, throat and nose form resonant systems. With the xylophone, tubes are fitted near the vibrating bars so that the air columns in the tubes will resonate and 'sound' when the bars are struck.

Loudspeaker Cabinets

A vented loudspeaker cabinet is another example. The air in the cabinet and the air in the vent or port form a resonant system which is driven by the sound radiating from the back of the speaker. At resonance, the air in the vent vibrates in phase with the direct radiation from the front of the speaker and so increases the output.

The cabinet is a version of a type of resonator which is very common in acoustics, the Helmholtz resonator, named after the nineteenth-century scientist who used them in his analysis of musical

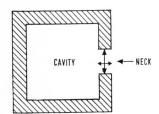

FIGURE 3.11

The basic essentials of a Helmholtz resonator are shown in the sketch. The mass of air (m) in the neck resonates with the compliance (Cm) of the enclosed air. The resonant frequency is given by:

$$f = \frac{1}{2\pi \sqrt{mCm}}$$

sounds. Many of the resonators in musical instruments depend on the Helmholtz principle and this is widely used in microphones to obtain good frequency response. Working in the reverse manner, Helmholtz resonators are used in the modern acoustic treatment of studios to control the reverberation by absorption.

Thinking of the loudspeaker cabinet once again, we see the essentials—an enclosed volume of air coupled to the free air by a neck or aperture. The volume of air in the neck is much smaller than the volume of the enclosed air. The enclosure can be any shape and of any material provided it is rigid and does not vibrate itself. If the enclosure does vibrate, then obviously the resonator action is modified.

The air in the neck acts like a piston which can compress the enclosed air. If we imagine a sound wave affecting the resonator initially, a compression, when it meets the neck air, will drive it inwards since the enclosed air will be at normal atmosphere pressure. The sound wave compression will be followed by a rarefaction and then the internal compressed air will be able to drive the neck air outwards. The neck air in these circumstances will vibrate to and fro and itself become a source of sound (see Fig. 3.11).

This will only occur when the frequency of the sound wave is the same as the natural resonant frequency of the resonator. We can see that, if the compressed enclosed air were trying to force the piston out before the sound wave compression had completely passed, the neck air would not be able to vibrate so far.

The frequency of the resonator depends on the volume of air in the neck and the volume of the enclosed air. Therefore, by varying these, we can alter the resonant frequency.

Of course the effect of a sounding board or resonator is not simply amplification of the sound. There can be a great deal of modification to the quality and the resonator can also affect the direction in which the instrument radiates its energy.

This is to be expected since, when driven and caused to vibrate by the original vibrations, resonators will impart their own characteristics to the final output. Their effect will depend on how broad or sharp their resonances are. With a broad resonance, a wide band of frequencies will be responded to, whereas with a sharp resonance, only sounds near the resonant frequency will be amplified.

CHARACTERISTICS OF SPEECH AND MUSIC

IN this chapter we are going to look at the characteristics of speech and music, what they consist of in terms of frequency and intensity, and see how these characteristics impose certain requirements on a recording or reproducing system. We shall also see how the various sources, the voice and musical instruments, distribute their sound in the space around them.

Speech

When listening to another person speaking, one is conscious that the voice has a characteristic pitch. A male voice has a lower pitch than a female and in turn the female voice is of a lower pitch than a child's. The average pitch for a man is about 130 c/s; that for a woman is about twice this. It is obvious that the actual pitch varies about these average values depending upon the individual.

The fluctuating air stream leaving the mouth has a complex waveform. As we saw in the last chapter, this will consist of a fundamental note along with its harmonics. It is the frequency of this fundamental which determines the pitch of the voice.

The cavities in the head resonate to accentuate particular groups of harmonics in the fluctuating air stream. These frequencies are called the formants. Varying the size of these cavities by moving the tongue, jaw, lips, etc., alters the resonant frequency for the various speech sounds.

Natural male speech has an overall frequency range of from 100–

8,000 c/s; for women the range is something like 200–10,000 c/s. Within these ranges there are some frequencies which are more important than others to the understanding of what is being said.

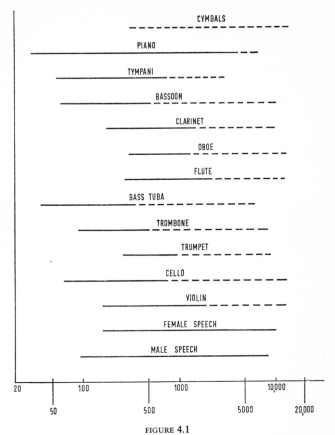

FIGURE 4.1

The frequency ranges of speech and some of the more common musical instruments. The solid line indicates the fundamental notes and the dotted extensions the harmonics.

Most of the energy in speech is contained in the low frequencies but they contribute very little to the understanding of speech. To put it another way, these frequencies do not control the 'intelligibility' of speech. When one hears only the low frequencies, say only up to 500 c/s, the speech sounds muffled and woolly and it is impossible to make any sense out of the speech.

It is the high frequencies which contain the intelligence. Listening to speech which has a severely restricted bass, one can still easily understand what is being said although of course it lacks power and sounds 'thin'.

Intelligibility

The question arises as to what range of frequencies is necessary just to understand speech. This question is obviously vital in the economical design of a telephone system, since the cost of a communication system depends on the frequency range—sometimes called bandwidth—it has to handle. With a telephone system, the main essential requirement is that the subscribers understand each other. This is examined by articulation tests, an articulation score of 100% meaning that the hearer is understanding perfectly every word being said.

High articulation scores can be obtained with a very narrow bandwidth, over 80% is possible with a range from 1,000–2,500 c/s. Of course with such a narrow frequency band the quality of the speech is poor, one can understand what is being said but perhaps not be able to identify who is saying it. Increasing the bandwidth improves the articulation score and this allows not only the intelligence but also some of the character of the speech to be transmitted.

A bandwidth of 2,700 c/s, ranging from 300 c/s to 3,000 c/s, is widely used in telephone systems. This gives reasonable economy but at the same time permits faithful reproduction of the important frequencies for intelligibility in the speech band. From a frequency range point of view, any high quality reproducing system of present-day standards should handle speech signals with fidelity.

As to volume, speech uses only a limited part of the intensity range of hearing. A volume range of some 40 dB will cover the requirements for faithful reproduction.

Directivity

Of course merely stating the frequency and intensity range of the human voice does not give us the whole picture. Another important factor is how the sound waves coming from the mouth are distributed in space. This information is given in a 'directivity pattern'. These are obtained by measuring the sound pressures produced at points which are at the same distance from the head, but at different angles. Using the pressure directly ahead at the measuring distance (two feet has been used in some investigations) as a reference, variations in readings at different angles show how the sound energy is distributed.

Taking the horizontal plane first, in front of the speaker the distribution is approximately the same for all frequencies over a total angle of some 120°. Over the other 240°, mainly behind the head, however, there is a considerable attenuation of the high frequency content. The low frequencies are only reduced slightly from directly

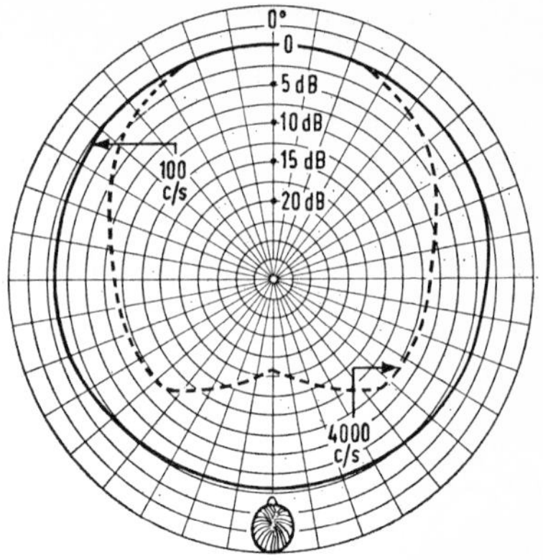

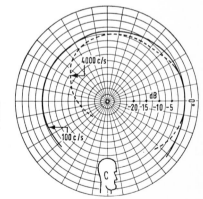

FIGURE 4.2

The directional characteristics of the human voice in the horizontal plane. (After Dunn & Farnsworth.)

in front of the head to directly behind. But at 5,000 c/s the pressure due to the sound wave is reduced by 20 dB. At 200 c/s the change is less than 5 dB (see Fig. 4.2).

In the vertical plane the distribution is again non-uniform at high frequencies. Starting at a point 45° below the horizontal, to right above the head, there is some divergence between the low frequency and high frequency distribution. Taking 5,000 c/s again, there is a

FIGURE 4.3

The directional characteristics of the human voice in the vertical plane. (After Dunn & Farnsworth.)

7 dB reduction; at 200 c/s the falling off in pressure is only 2 dB right above the head. Behind the head, as we would now expect, the high frequencies are attenuated much more severely than the low frequencies (see Fig. 4.3).

64

This variation in directivity of the voice is obviously important in the microphone technique used in recording or broadcasting. For good reproduction, the microphone should be not too far away from a line directly in front of the head.

Music

Music has a much wider frequency and intensity range than speech. From a frequency point of view, a high fidelity reproduction system for music should be capable of handling a range from 30 c/s to 15,000 c/s. The maximum intensity range of music directly heard is of

TABLE 4.1. *Energy Levels of Musical Sounds*

Origin of Sound	Energy
	Watts
Orchestra of seventy-five performers, at loudest	70
Brass drum at loudest	25
Pipe Organ at loudest	13
Trombone at loudest	6
Piano at loudest	0·4
Trumpet at loudest	0·3
Orchestra of seventy-five performers, at average	0·09
Piccolo at loudest	0·08
Clarinet at loudest	0·05
Human voice ⎰ Bass singing ff	0·03
Alto singing pp	0·001
Average speaking voice	0·000024
Violin at softest used in a concert	0·0000038

(Courtesy Bell Telephone Laboratories.)

the order of 70 dB. This is too wide for most recording or broadcasting systems and some compression is necessary.

The frequency range just quoted is of course for the entire range of sounds produced by the musical instruments which are normally used. The range can be broken down, first of all into the individual ranges of the various instruments and then, for a particular instrument, into the range of the fundamental notes and overtones it produces.

Most musical notes are complex, consisting of a fundamental and a series of harmonics. With the oboe, the range of the fundamental notes it produces is from 233 c/s to 1,574 c/s approximately. The harmonics extend this range to about 11,000 c/s.

The violin has a fundamental range of about 1,000 c/s starting with its lowest string at 200 c/s. With the harmonics, the highest frequency

E

produced is round about 14,000 c/s. It is the same with all the sources of musical sounds, the human voice and the instruments. And to hear these sources properly, not only must all the frequencies—funda-

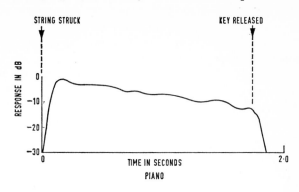

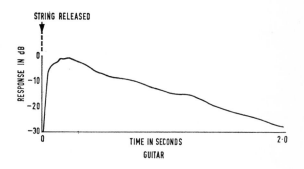

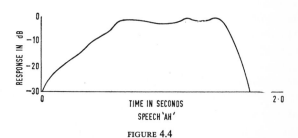

FIGURE 4.4

These graphs show how the growth and decay of sound varies for three types of source. (After Olson, *Musical Engineering*, McGraw-Hill Ltd.)

mentals and harmonics—be present, but they must retain their correct amplitudes relative to each other.

So much for frequency; from an intensity range point of view, two

66

factors are important. The maximum range, which we have already seen is 70 dB, is from pianissimo to fortissimo. In practice, much music does not exploit this full intensity range; the range depends on how the individual composer has used the instruments at his disposal.

Variation with Time

The other factor is how the intensity varies with time; the manner in which the sound builds up, how it behaves when it has reached its maximum intensity and how it decays. These qualities depend on the type of instrument and the player. The construction will control how the energy applied to the instrument by the player is utilised and, of course, the player himself also determines the intensity characteristics by his playing.

With an instrument such as the clarinet, both the build-up and the decay are fairly short and it sustains a steady output at its maximum value.

A stringed-instrument like the harp, which is played by plucking, builds up its intensity quickly but it cannot sustain the note at a steady intensity. The decay starts immediately the maximum has been reached. The strings are lightly damped and this means they sound for some considerable time, giving a long decay.

The piano is the most common example of a struck string instrument. Here the intensity pattern is basically the same as with the plucked string instruments except that the decay period can be longer. Of course this is considerably reduced by the damping action when either the key is released or the damping pedal is operated.

A bowed-string instrument—such as the violin—has a rapid rise to the maximum intensity, which can be sustained. The decay time can be relatively long or it can be short depending on whether the string is left to vibrate freely or whether it is touched. Another example of a struck instrument is the drum, which produces a rapid rise in intensity. The note is not sustained and the decay time is relatively short.

With the voice, the intensity characteristics vary considerably but generally the build-up is longer than with most musical instruments. It can sustain a note at a steady intensity and the decay is short.

Directional Characteristics

We have seen that the spatial distribution of the voice's energy forms a complicated pattern depending on both angle and frequency. At low frequencies—long wavelengths—the distribution is almost uniform. With high frequencies—smaller wavelengths—the sound becomes concentrated over a smaller frontal region.

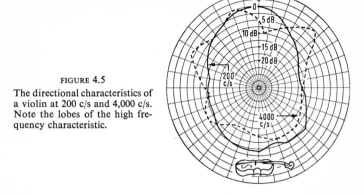

FIGURE 4.5

The directional characteristics of a violin at 200 c/s and 4,000 c/s. Note the lobes of the high frequency characteristic.

The same general distribution pattern is produced by musical instruments. When the wavelength is large compared with the source dimensions, there is very little concentration or beaming of energy. As the frequency rises the wavelength gets smaller. When it approaches or gets smaller than the source dimensions, the distribution becomes directional, markedly so in some cases.

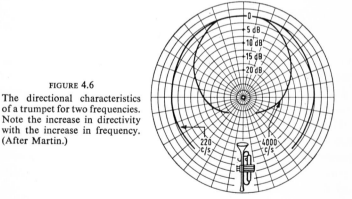

FIGURE 4.6

The directional characteristics of a trumpet for two frequencies. Note the increase in directivity with the increase in frequency. (After Martin.)

The complicated construction of some instruments cause their directivity patterns to have many lobes. These can vary rapidly in direction and intensity. They are due to interference between the sound waves coming from more than one source of sound. With the woodwind for example, clarinet, oboe, etc., sound comes from the holes as well as the mouth. The front and back of the violin body both vibrate over a wide range of frequencies and this produces lobes. However, with brass instruments, such as the trumpet and trombone, the sound comes only from the mouth and the distribution is much smoother.

Musical Instruments—Their Principle of Operation

In this section we are going to examine the various types of instruments and see how they operate, using the basic principles outlined in Chapter 3. There we saw that each instrument consists essentially of a resonator which is excited by either blowing, striking, bowing or plucking. The resonator may be an air column, a string, a plate, a bar, or a stretched membrane. This vibrates at certain frequencies

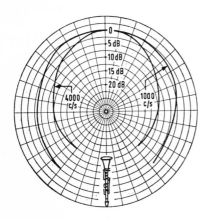

FIGURE 4.7
The directional characteristics of a clarinet. The increase in directivity at the higher frequencies can be appreciated from the sketch. It should be noted that there are small sharp lobes in the directivity pattern due to radiation from the holes, but these have been smoothed out. (From Olson, *Musical Engineering*, McGraw-Hill Ltd.)

and, in some cases, its acoustical output is amplified by coupling to it a secondary vibrating body or sound board.

The instruments using these resonators are, of course, normally given the family names of woodwind, brass, strings and percussion.

Woodwind and Brass Instruments. Although all woodwind and brass instruments use an air column, there are several methods used in causing the column to vibrate. It is first of all necessary to produce a pulsating or fluctuating air stream from the steady air expelled from the mouth.

With the woodwind, the pulsations are produced either by the air itself, used in a certain manner, or by a mechanical reed. With the brass instruments it is the player's lips which vibrate.

The flute is an example of the first method. It consists of a cylindrical pipe closed at one end and open at the other. Near the closed end is the blowhole—technically called the 'embouchure'. Air coming

69

from the player's mouth strikes the wedge-shaped edge of the embouchure. When an air stream strikes a sharp edge in this way, small pockets of air bounce alternately from one side and then the other. The note produced, due to the air forming eddies, is called an 'edge tone'. These edge tones provide the force necessary to set the attached air column into vibration.

Mechanical reeds are divided into two classes, single and double. The clarinet is an example of a single reed instrument, while the oboe uses a double reed.

Taking the single reed first, this is clamped at one end by a grip. The free end lies across the opening in the mouthpiece of the instrument. When the player blows, differences in pressure between the two faces of the reed are produced which cause it to vibrate. Naturally there is a small gap between the reed and the opening, so that air can enter the opening.

When the steady air stream strikes the reed some air will be able to flow into the instrument down one side of the reed. On the other side there is comparatively little air flow. This means that there is a difference in the air velocity down the two faces of the reed, which produces a pressure difference. In fact the pressure on the inside face is less than the outside, so that the reed moves in to close the gap. However, when the gap is reduced in size, the air flow into the instrument becomes restricted and the pressure in the inside face rises above that on the outside face. This causes the reed to move in the other direction. In moving, of course, it increases the size of the opening, the air flow into the instrument also increases and hence the pressure on the inside face once again falls. The reed moves inwards to close the gap and the cycle is repeated over again. From this we see that the air stream injected into the instrument is continually pulsating.

With the double reed, the action is basically the same except that there are two vibrating surfaces. Air enters the instrument through the space between them and once again pressure differences between the air in this space and outside the reed cause them to vibrate.

It is possible to regard the player's lips as a double reed in the action of brass instruments. The steady air stream from the breath strikes the lips, which are surrounded by the mouthpiece of the instrument. Pressure differences between the inside and outside of the lips cause them to vibrate and this interrupts the air flow in a regular manner.

Once there is a pulsating air stream, the air columns are set into vibration and they then mainly dictate the pitch of the note the instru-

70

ment produces. With brass instruments, adjustment of lip tension and blowing pressure can also alter the resonant frequency.

We saw in Chapter 3 how the length of the column decides the pitch and with the orchestral instruments of the woodwind and brass families, the players can vary the effective length of the air column to produce a wide range of notes.

On woodwind instruments such as the flute, clarinet and oboe, there are a series of holes down the instrument. These can be covered either by the fingers or with valves operated by keys. Opening and closing the various holes produce different effective lengths of air column and so set up the required resonant frequencies.

In modern woodwind instruments there is a complicated system of links and shafts which allow the player to open—or close—more than one hole at a time. This gives the player more control over the resonant frequency of the air column and enables him to alter the air length more widely and easily than he could if he had to rely on his fingers alone.

Brass instruments vary the length of air column either by adding lengths of side tubes or by telescoping two tubes together. The trumpet has three valves which allow various combinations of extra lengths to be introduced into the air column. The French horn, cornet and tuba also use valves. The length of the column in the trombone is varied by the slide or telescopic section. This gives continuously variable control over the resonant frequency, unlike the other members of the brass family which can only change the length by certain fixed amounts. This allows the trombone player to produce gliding effects as with the violin.

Ranges of the Woodwind and Brass Instruments

Woodwind. The piccolo is the smallest woodwind instrument used in the orchestra. It is a smaller version of the flute, held in the same way across the mouth, and works on the same principles. The fundamental frequency range is about three octaves. The length is just over one foot.

The flute has the same range as the piccolo—three octaves—but it is pitched an octave lower than the piccolo. Its length is about twenty-seven inches.

The fundamental notes of the clarinet have a range of over three octaves. The clarinet tends to behave like a quarter-wavelength pipe and its characteristic tone is due to the odd harmonics being accentuated (see p. 49). The even harmonics are actually present but are much weaker. The length is just over two feet. There is also

71

the bass clarinet, a larger instrument which is pitched one octave lower.

The oboe, as we have seen, is a double-reed instrument with a range of about three octaves. In the concert hall the oboe normally sounds the note A as a tuning note to which the rest of the orchestra tune their instruments. All the even and odd harmonics are produced, the column acting like a half-wavelength pipe. The oboe does not have a flared bell like the clarinet. The instrument measures about two feet.

The English horn or *Cor Anglais* is similar to the oboe except that the air column is terminated with a spherical bulb. Its range is less than that of the oboe and it is pitched lower. The length is about three feet.

The last of the woodwind we shall consider is the bassoon, another double-reed instrument working with a long air column about eight feet in length. To make it easier to handle, the tube is folded. This length means the pitch is low, two octaves below the oboe. The range is about three octaves.

Brass. The trumpet has a pipe length of about six feet. The bore is not uniform throughout its length, being partly cylindrical and partly conical. The mouthpiece is cup-shaped and the open end bell-shaped. We have already seen how the length of the pipe is altered by the valves. To produce a lower output of sound and to alter the tone quality, a 'mute' may be inserted into the bell. This is made of either plastic or metal. The trumpet has a range of about three octaves and it measures about two feet.

The French horn is an unusual instrument in that the player inserts his hand into the large bell-shaped mouth. By doing this he is able to modify the length of the air column slightly and alter the pitch. It also allows him to produce muted effects. The physical length of the tube is about twelve feet. It is coiled and fitted with valves. With these valves, his hand in the bell and his lip tension the player is able to control the pitch over a range of over three octaves.

We have already discussed the trombone, seeing how the telescope sections alter the effective length of the air column. The total length of the tube is about nine feet and the instrument has a range of about two and a half octaves in the smaller of the two versions available. This is called the tenor trombone. The bass trombone, being larger, has a lower pitch and covers another half-an-octave in range, making three in all.

The tuba is another brass instrument with valves. The actual length of the tube is approximately eighteen feet; this is coiled and ends in a large bell-shaped mouth. As with most brass instruments, the fundamental notes it produces range over about three octaves.

Stringed Instruments

We have seen in Chapter 3 how a string vibrates to produce a fundamental frequency along with its harmonics. This gives the sound produced by the string a certain 'timbre' or tone quality, and this quality depends on where and how the string is touched. To produce a sound, the string must be displaced from its rest position and to do this three methods can be used—plucking, bowing and striking. These different methods of excitation produce different tone qualities.

Most instruments use only one method but there are some which, although using one main method, do allow all three to be used. The use of soundboards and other resonating devices to amplify the very weak sounds from vibrating strings has been mentioned previously.

Plucked-String Instruments. In this category come the harp, guitar, mandolin, banjo and harpsichord. The harp has a vertical string arrangement, these being stretched on a frame. This frame is made up of three principal members, pillar, neck and soundboard. The strings are stretched between the neck at the top and the soundboard at the bottom. The strings vary in length, the short ones being near the end where the neck and the soundboard join, the long ones at the end where the pillar separates the neck and soundboard. The instrument has a height of almost six feet. The player uses his fingers to pluck the strings, and of course the various string lengths produce the different notes. But there are also seven pedals which allow the player to shorten the effective length of the strings and raise the pitch. The range of the fundamental notes is six and a half octaves.

The guitar, mandolin and banjo have several features which distinguish them individually. The guitar has a fairly large body. There are (usually) six strings stretched between pegs and a tailpiece. The mandolin has a set of eight strings arranged in four pairs, stretched between pegs and over a bridge. The body has a flat top but is curved at the back. The banjo has four long strings and a short one, stretched in the same way as the mandolin. The bridge is attached to a membrane which acts as one end of a cylinder formed by the case— the other end is open.

However, all three instruments have one common feature, the strings are stretched over a fretted finger board. The frets are raised ribs placed across the board. To vary the pitch of the note, the string is pressed against the fret. The distance between the frets is arranged so that the notes produced differ in pitch by a set amount. The player uses a flat piece of plastic or metal called a plectrum to excite the strings. With the guitar and the banjo, the player sometimes uses his fingers instead of a plectrum. The guitar and banjo measure about three feet while the mandolin is about two feet. The instruments each have a range of about three octaves.

The harpsichord has a larger number of strings, stretched across a frame, which are excited by plectrums. These are actuated from a keyboard, rather like that on a piano. The fundamentals produced by the instrument range over approximately four and a half octaves.

Bowed-String Instruments. The violin family—violin, viola, cello and double-bass, come into this category. They are normally played by drawing a bow, made of horsehair stretched between two pieces of wood, across the strings. The strings, stretched between a tailpiece and the pegs, pass over the bridge and run just above the finger board. Since they are under tension, there is a restoring force bringing the strings back when they are displaced from their rest position. The bow pulls the string away from the rest position until the restoring force becomes sufficient to bring it back. Then the bow grips it again and pulls it away. The string is thus drawn into vibration and will continue to vibrate and sound a note as long as the bow movement is maintained.

Of course the string can be made to vibrate by other means—by plucking with the fingers or by striking the strings with the wooden back of the bow. The difference between the three methods is in the tone colour of the sounds produced.

When discussing the trombone, we saw that the length of the column could be varied continuously and not in discrete steps like the other brass instruments which use valves. There is exactly the same comparison between the violin family and the fretted instruments like the guitar, mandolin, etc. This allows the player much greater freedom in obtaining and controlling any desired note.

The four members of the violin family are very similar in construction. They of course differ in size, the violin is about two feet in length, the viola is slightly larger, the cello measures just over four feet and the double-bass about six and a half feet. Since the size determines the frequency range, the fundamental frequency ranges of the instruments are different. However, the ranges are about the same in extent, about three octaves.

Percussion

This class contains a large number of instruments: drums of various shapes and sizes, cymbals, bells, xylophone, triangle, tambourine.

These are just some examples and it is interesting to note that they have one feature in common: the vibrating system is excited by being struck. Sticks, wooden mallets or metal rods are used. The vibrating system, as we saw in Chapter 3, can be a membrane, a bar, a plate or a bell.

Percussion instruments can be divided into two categories, those having a definite pitch and those which really produce a noise which has no definite pitch. Examples of definite pitch instruments are the tympani, xylophone and bells. The bass drum, cymbals and triangle are three examples of instruments having indefinite pitch.

Piano

This is a struck string instrument and is sometimes included in the list of percussion instruments. The metal strings are stretched across a steel frame and they are struck by felt covered hammers. The vibrations of the strings are coupled to a soundboard which, as we have seen before, amplifies the vibrations and so increases the acoustic output.

The hammers are caused to move by keys; there are 88 keys on a standard piano and each key has a string, or group of strings, associated with it. Pedals are provided which allow the notes to be either sustained or damped. The piano has a very wide frequency range extending over seven octaves.

Musical Scales

In Chapter 1, we saw that pitch was the subjective quality of a musical note which enabled one to place it on the musical scale. To conclude this chapter we are going to look briefly at musical scales and see how they are built up.

Most people are very familiar with the names of the notes in the tonic sol-fa system—if only from school days. (d) Doh, (r) re, (m) me, (f) fa, (s) so, (l) la, (t) te, (d) doh are the well-known notes of the major scale. Most people recognise, probably by picking the notes out on a piano, that in going from d to d' we have gone through an octave—eight notes. Increasing the pitch by an octave corresponds to a doubling of frequency, the frequency of d' being twice that of d.

The octave is a ratio between two notes which produces a pleasant sensation, the notes are said to be in consonance, or they harmonise. An unpleasant sensation is produced when two notes sounded together are in discord or are said to be dissonant.

There are other ratios besides 2 : 1 which harmonise, and the octave can be split up into several such ratios. These form the basis of a scale which has been used in Western music for centuries.

It is called the Major Diatonic Scale and it is based on the fact that the absence of dissonance when two notes are sounded together depends on whether the notes have a ratio based on simple factors. Other scales are, of course, possible. Pythagoras for example used only two types of note interval when dividing the octave. The major diatonic has probably some connection with the Pythagorean scale. There is also a five note, or pentatonic scale, used in bagpipe music for example. The Arabs divide their scale into sixteen ratios, while the Hindus divide theirs into twenty-two.

Musical Intervals

It is interesting to note that our appreciation of pitch changes is similar to the way in which the ear reacts to intensity changes. In Chapter 1 we saw that the significance of a change of intensity depends on the *ratio* of the intensities. Likewise a pitch change depends on the ratio of the frequency of the two notes. An octave change is obtained by doubling the frequency, irrespective of what the original frequency was. For example, going from 128 c/s to 256 c/s and then to 512 c/s gives two successive changes of an octave, irrespective of the fact that the arithmetic difference between 512 and 256 is greater than that between 256 and 128. This confirms that the effect of pitch changes depends on ratios and not subtracting or adding cycles per second. The ratios are termed intervals.

Now we can build up the list of Table 4.2, starting with unison (1 : 1) and finishing at the octave (2 : 1), to include the common ratios.

TABLE 4.2. *List of Musical Intervals*

Ratio	
1 : 1	Unison
16 : 15	Semitone
10 : 9	Minor Tone
9 : 8	Major Tone
6 : 5	Minor Third
5 : 4	Major Third
4 : 3	Perfect Fourth
3 : 2	Perfect Fifth
8 : 5	Minor Sixth
5 : 3	Major Sixth
9 : 5	Minor Seventh
15 : 8	Major Seventh
2 : 1	Octave

Taking our tonic sol-fa again, we can see how these ratios and terms are arrived at. Table 4.3 shows each of the sol-fa notes with the usual notation and gives the starting, or key-note, a frequency '*f*'. This makes the expressions general and so lets us see the pattern of semitones and other intervals.

TABLE 4.3. *Frequency Ratios in Major Scale*

Tonic Sol-fa	d	r	m	f	s	l	t	d′
Notation	C	D	E	F	G	A	B	C
Frequency	f	$\frac{9}{8}f$	$\frac{5}{4}f$	$\frac{4}{3}f$	$\frac{3}{2}f$	$\frac{5}{3}f$	$\frac{15}{8}f$	$2f$
Interval Ratios		$\frac{9}{8}$	$\frac{10}{9}$	$\frac{16}{15}$	$\frac{9}{8}$	$\frac{10}{9}$	$\frac{9}{8}$	$\frac{16}{15}$

The interval ratios we can see are of three kinds:

$\frac{9}{8}$ called major tones

$\frac{10}{9}$ called minor tones

$\frac{16}{15}$ called semitones

If we go from the first note C to the *third* note E, the ratio is 5 : 4, and Table 4.1 shows that this is a major *third*. If we go from C to the *fourth* note F, the ratio is 4 : 3 and this is called a perfect *fourth*. Similarly, going from C to A gives a ratio of 5 : 3—a major *sixth*.

Notice that the scale consists of a key-note, major tone, major third, perfect fourth, perfect fifth, major sixth and major seventh. This is the major scale, based on a key-note of C. A minor scale is also used in which the intervals produce minor thirds, minor sixth and minor seventh, i.e. the ratio between the third note and the key-note is 6 : 5 and so on. This would occur if we, for example, chose the note E as our key-note and played on the white notes of the piano. Going from E to G would provide a ratio of $\frac{3}{2} \times \frac{4}{5} = \frac{12}{10} = 6:5$—a minor third as we have seen.

These two scales are called scales of 'just intonation'. They produce ratios which are acceptable as pleasant changes in pitch. However, they suffer from a practical disadvantage with certain types of musical instruments.

To see how this arises, let us take the scale of C major and compare it with one using a different key-note, say E. This time we will insert frequencies instead of ratios and we will take the frequency of A as 440 c/s. This is the International Concert Pitch.

77

Table 4.4 shows that for some of the notes the frequency has changed. G for example, is not the same in the E scale as it was in the C scale. In music it is often necessary to change the key-note—to *modulate* is the usual term. But as we have seen from the Table,

TABLE 4.4. *Limitation of Just Intonation Scale*

	C	D	E	F	G	A	B	C	D	E
Key of C	264	297	330	350	396	440	495	528		
Key of E			330	371·25	412·5	440	495	550	618·7	660

changing the key changes the frequency of some of the notes. With the voice, or the violin, these changes can be offset. The singer can adjust the pitch of the note until it coincides with the new key-note and the violinist can adjust the length of string.

But with fixed tone instruments, the piano for example, such adjustment is impossible and this means that an impossibly large number of tones would be required to cover all the possible changes of key. Each octave would have to include a substantial number of additional notes.

The Equal Temperament Scale

One can therefore see that the scale of just intonation has practical difficulties. Because of these, another scale has evolved in which the intervals are of the same value for all key-notes. Obviously this means that the scale is a compromise; only the octave interval is accurate, while the others are slightly different from that demanded by pure consonance. However, this discrepancy is unavoidable if the practical difficulties of the just intonation scale are to be overcome.

With the octave divided into equal intervals, this scale is known as the Equal Temperament Scale and there are twelve equal semitone intervals to each octave. If we call this interval x for the moment and, for simplicity, give our key-note the value of 1, we can calculate the interval ratio as in Table 4.5.

TABLE 4.5. *Equal Temperament Intervals*

	Key-note 1	2nd note x	3rd note x^2	4th note x^3	5th note x^4
Interval	$\dfrac{x}{1} = x$	$\dfrac{x^2}{x} = x$	$\dfrac{x^3}{x^2} = x$	$\dfrac{x^4}{x^3} = x$	

The thirteenth note will be x^{12} and, since this will be the octave, it follows that:

$$x^{12} = 2$$

or $$x = 2^{\frac{1}{12}} = 1 \cdot 0594.$$

With such a ratio it is obvious that there will be some discrepancy between notes of the just intonation scale and those on the scale of equal temperament. Some notes on the latter scale are slightly high in frequency or *sharp* and some slightly low or *flat*, compared to their corresponding notes in the former scale.

HOW MICROPHONES WORK

ALL microphones work on the general basic principle that the energy of the sound wave is converted firstly into mechanical energy and then into electrical energy. So they all need a diaphragm which will vibrate when the sound wave produces a difference of pressure between its faces, and some means whereby mechanical movement can cause electrical signals to be generated.

It is interesting to note that this is opposite to the action of a loudspeaker, where electrical energy is fed in to produce the mechanical vibrations which cause sound waves to be set up in the air. The microphone is an example of an electrical generator, while the loudspeaker is an example of an electrical motor.

Methods of Generating the Electrical Output

Although it might seem that we are starting at the wrong end of the chain, it is perhaps easier to discuss the ways of generating current microphones because most people refer to microphones as being 'moving coil', 'ribbon', 'crystal', 'condenser', etc. These terms of course refer to the methods by which the mechanical energy is converted into electrical energy. Other methods are available, but these are the ones normally used in recording and broadcasting.

Carbon. An example of one of the other methods is the carbon microphone which is used in telephones. Carbon granules are packed into a box with the diaphragm as one side. When the diaphragm vibrates, the granules are subjected to a varying pressure, so that the areas of contact between the granules alters and the resistance of the device to electrical current will change. Thus, if a current is passed

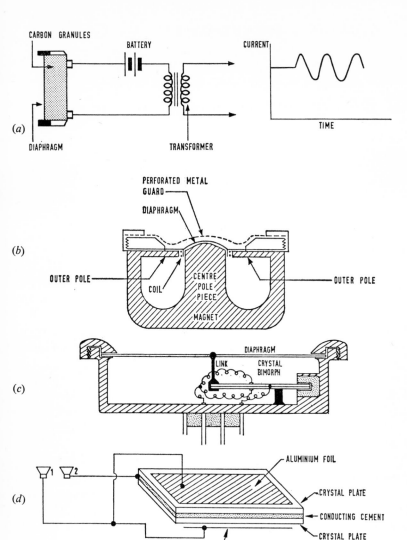

FIGURE 5.1

(a) This shows the essentials of a carbon microphone and how it is connected. The battery causes a uni-directional current to flow round the circuit. The current fluctuates about this steady value, and the transformer passes only the fluctuations and isolates the steady current.

(b) A typical moving coil microphone.

(c) In one type of crystal microphone the diaphragm is attached at its centre to one end of a bender bimorph of Rochelle Salt.

(d) Showing the bimorph connections to terminals 1 and 2. (From *Disc Recording and Reproduction*, by P. J. Guy.)

F

81

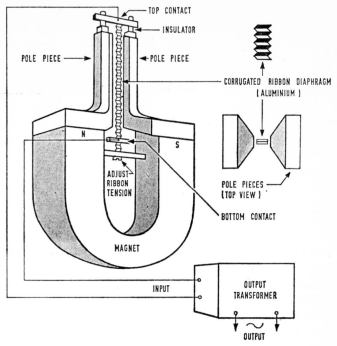

FIGURE 5.2

This shows the basic arrangements inside a modern high quality ribbon microphone in which the magnet is mounted below the pole-pieces. The ribbon, made of very thin pure aluminium foil, is corrugated to give flexibility and to prevent curling. The resistance of the ribbon is less than 1 ohm and a transformer is built in to give the required impedance.

The small sketch shows how the pole-pieces are shaped to provide a cavity on each side of the ribbon. These cavities are broadly resonant to ensure that the high frequency response is maintained.

through the carbon granules from a battery, the varying resistance will cause variations in the current and these variations are an electrical replica of the variations in pressure in the sound wave. The carbon microphone was used in the early days of recording and broadcasting but its quality is now no longer acceptable. However, for telephony its cheapness, robustness and performance are adequate.

Moving Coil and Ribbon. These two microphone types use the same basic idea—if a conductor moves in a magnetic field a voltage will be produced across its ends. This is the fundamental idea behind a bicycle-lamp dynamo, the dynamo in a motor car or the generator at a power station. In large generators it is usually more practical if the conductor remains stationary while the magnetic field revolves. In microphones, however, the magnetic field is provided by a fixed permanent magnet which has pieces of soft iron known as pole pieces attached to it so that an intense field is produced across a narrow gap. The magnitude of the output voltage depends on the strength of the field, and this should obviously be as high as possible.

In the moving coil microphone, the gap is an annular one and the coil is held in position by the diaphragm. When the diaphragm vibrates, the coil moves in the magnetic field and a voltage is produced. The frequency and amplitude of this voltage will be controlled by the frequency and amplitude of the sound wave. The moving coil type is often referred to as a 'dynamic' microphone.

In the ribbon microphone, the conductor and diaphragm are one and the same. It consists of a thin strip of metallic foil placed between two elongated pole-pieces (see Fig. 5.2).

Crystal. If certain materials are mechanically deformed, i.e. bent or twisted, a difference of voltage is produced between the faces of the material. This is known as the piezo-electric effect and is obviously applicable to microphones since we can use the forces due to a sound wave to drive a thin layer of suitable material. Although there are several possible materials, the one normally used is Rochelle Salt since this is very sensitive and produces a good output signal.

A typical arrangement is one in which two thin layers of Rochelle Salt are sprayed with metallic coatings to provide the electrical connections. The layers are fixed together back to back so that the two

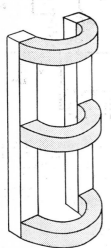

FIGURE 5.3
In some ribbon microphones the magnetic field is supplied by three magnets placed across the pole-pieces as shown in the sketch.

faces in contact form one side of the generator and the exposed faces are connected together to form the other side. The layers are fixed at three corners and the fourth corner is attached to the diaphragm.

When the diaphragm vibrates, the crystal layers are twisted and a voltage is produced which again will be controlled by the frequency and amplitude of the sound wave (see Fig. 5.1c).

There is another arrangement in which two separate crystal units are directly actuated by the sound waves.

It must be noted that the piezo-electric effect of a material depends on humidity and temperature. The material used is generally given a protective coating to withstand the effects of humidity, but temperatures in excess of 45° C (113° F) will destroy the crystal.

Condenser. Here the diaphragm is one plate of a condenser, the other plate being fixed and the condenser being kept in a state of charge by a d.c. supply. When the diaphragm vibrates, the distance between the plates varies and hence the capacitance will vary. Now if the charge on a condenser can be kept constant while its capacitance varies, the voltage across the plates will vary in sympathy. Since the variations in capacitance are due to the sound wave, the voltage output will follow the variations of the sound wave. The necessary com-

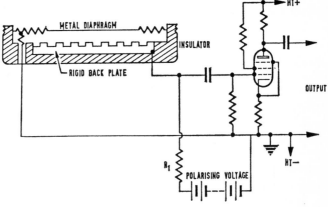

FIGURE 5.4

The essentials of a condenser microphone. The polarising supply, of the order of 60 volts, charges the capacitance between the diaphragm and fixed plate. An alternating voltage is produced when the capacitance is varied by the diaphragm vibrating. The load resistor, of very high value, ensures that the charge remains substantially constant. A blocking capacitor isolates the D.C. of the polarising supply from the head amplifier.

ponents are the condenser, or 'capsule' as it often is referred to, a charging or polarising d.c. supply, and a load resistor. These are connected in series, the load resistor having a very high value—a representative value would be 250 MΩ—to ensure that the charge remains reasonably constant (see Fig. 5.4).

In practice the diaphragm is made of metallic foil or metallised plastic about 0·001 in. thick. This is mounted very close to the fixed plate so that when the diaphragm vibrates there is a large change in

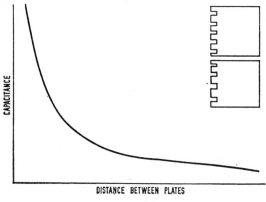

FIGURE 5.5

The capacitance depends on the distance between the plates and it can be seen that the largest change in capacitance will be produced when the initial distance is small. With small distances, however, the trapped air between the diaphragm and fixed plate would be so stiff that the diaphragm would be unable to vibrate. To increase the amount of air without decreasing the effective capacitance, the fixed plate is grooved or slotted as shown above. The fixed plate is perforated by a pressure equalising tube for pressure operation.

capacitance. Clearances of the order of 0·001 in. are used. The capsule behaves like a capacitance of between 20 and 60 pF. A typical value of polarising voltage is 60 V.

Methods of Driving the Diaphragm

The diaphragm will vibrate when a *difference* of pressure exists between its faces and obviously one way to produce this difference would be to design the acoustical system of the microphone along the same lines as the human ear.

In Chapter 2 we saw that the ear-drum vibrates because one of its faces is enclosed by the middle ear, the air pressure in this part of the ear being kept at the steady atmosphere value. The other face is exposed to the sound wave, which is a variation in air pressure about the steady atmosphere value, and so the ear-drum vibrates.

Exactly the same principle is applied to microphones, the system of operation being called *pressure operation*. The inner face of the diaphragm is enclosed by an air chamber which has a pressure equalising tube functioning in the same way as the Eustachian tube in

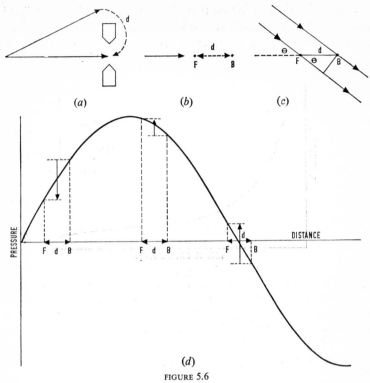

FIGURE 5.6

Pressure-Gradient Operation. A driving force is produced by the difference in pressure between the faces of the ribbon due to the existence of the path length 'd'.

(a) The extra distance which the sound wave has to travel to reach the back of the ribbon is shown and the effect on 'd' of the dimensions of the pole-pieces or any surrounding shields can be seen.

(b) As discussed in the text, the front (F) and back (B) of the ribbon can be represented as two points at which the sound wave is sampled. They are separated by 'd'.

(c) For a sound approaching from angle θ the effective path length is FX which is $d \cos \theta$.

(d) Using a Pressure/Distance graph one can see how the driving force is made to vary in direction and amplitude by the sound wave. Since the path length is a distance, it can be represented along the distance axis.

Of course, in practice the microphone stands still and the sound wave passes it. In drawing, however, it is convenient to keep the wave still and move 'd' along the distance axis. Each drawing of 'd' in the sketch represents what is happening at different times when the wave passes the microphone.

At the first time the pressure at B is greater than at F, this is indicated by the direction of the arrow.

At the second time the pressure at B is now less than that at F, the driving force has now changed direction.

At the third time the driving force has increased to its maximum value. Increasing the amplitude or the frequency will obviously increase the driving force.

the ear. The outer face is open to the sound waves, and thus the diaphragm will vibrate in sympathy with the frequency and amplitude of the sound wave.

Pressure Gradient Operation

Another way to produce the necessary difference of pressure between the faces of the diaphragm is to allow the sound wave access to both sides of the diaphragm, but at different times. Instead of comparing the pressure variations of the sound waves with the steady atmosphere pressure, the pressures at two separated points in space are compared and their difference causes the diaphragm to vibrate.

This is called *pressure gradient operation*. Let us call our two points *A* and *B*, and examine what happens just as the first compression is passing *A*. Since the points are separated by a finite distance, it will take a little time for the compression to reach *B* and we can see that at any instant there will be a pressure difference between points *A* and *B*. In fact the difference between them is controlled by the frequency and amplitude of the sound wave (see Fig. 5.6).

Another way of explaining this is to say that there is a *difference in path length* from the source to *A* and from the source to *B*. And so long as there is a difference in path length there must be difference in pressure, provided certain conditions are met. We shall see later what these conditions are. Instead of referring to the 'difference in path length', we can refer to the 'path length' and give it the symbol '*d*'. It might at first be thought that *d* is due simply to the thickness of the diaphragm, but diaphragms have to be very thin—sometimes as little as 0·000025 in.—so that they will vibrate easily. Consequently the difference in pressure would not be very much. In fact *d* is dependent on what surrounds the diaphragm. In a ribbon microphone it is the way in which the pole-pieces, and sometimes acoustic shields, are arranged; in condenser microphones it depends on how the fixed plate is perforated and how the case containing the capsule is arranged.

In the previous paragraph it was stated that certain conditions must apply before we get a difference in pressure, when we have a difference in path length. Since the path length is a physical distance, there must be some frequency for which *d* is exactly equal to a wavelength. Remembering that the wavelength is defined as the distance between two layers of air which are experiencing the same pressure conditions at the same instant of time, we can see at once that when $d = \lambda$ there will be no difference in pressure due to *d*.

This is termed the first extinction frequency. At half this frequency,

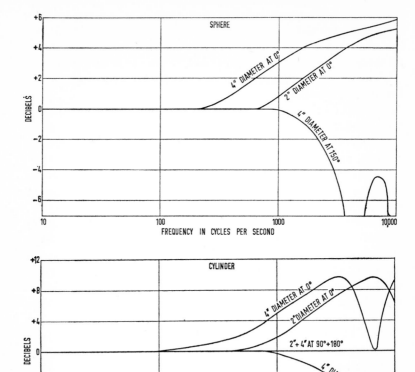

FIGURE 5.7

When an object is placed in the path of a wave any effect it may have
depends on four factors: the size of the object, the shape of the
object, the frequency of the sound wave, the angle of incidence.
The above graphs show how the pressure varies when these factors
vary. The free field pressure P_0, i.e. with no object present, is first
measured. Then, with the object in position, the pressure P at a
point on the object is measured. The ratio in decibels is then obtained
from $20 \log_{10} \dfrac{P}{P_0}$. With a sphere the point can be anywhere on the
surface; with a cylinder it is on one of the faces.
The graphs are drawn for spheres and cylinders of 2 in. and 4 in.
diameter to show the effect of size and shape. The 150° angle of
incidence was chosen since at this angle, in the main, the variation at
high frequencies is the widest.

the difference in pressure is a maximum since *d* is half-a-wavelength. From this we can see that, with a fixed *d*, the difference in pressure between the faces of the diaphragm depends on frequency. In practice this must obviously be subsequently corrected, and it is done by careful design of the mechanical system and layout of the microphone.

Directivity Patterns
The graph of a microphone's response to sounds coming from different angles is called a polar diagram or directivity pattern. This is obviously important as, in most cases, sounds reach the microphone not only by the direct path but by reflected paths from the walls of the room, etc. The shape of its directivity pattern has a great bearing on how a microphone should be placed so that an acceptable balance between direct and reflected sound is obtained.

Another useful point is that a microphone with known directional properties can be angled in such a way that the contribution from an undesired source can be considerably attenuated.

Ideally, of course, the directivity pattern, whatever its shape, should be the same at all frequencies in the working range of the microphone. However, in practice this is somewhat difficult to achieve because of the large range of wavelengths a microphone has to handle. As a result very few microphones have directivity patterns which are completely independent of frequency.

Omni-Directional. The simplest pattern is of course a circle, indicating that the microphone is equally sensitive in all directions—the so-called *omni-directional* pattern.

This is obtained by having a pressure operated microphone. As we are simply comparing the sound wave pressure with that of the atmosphere, it does not matter from what direction the sound wave approaches the microphone. From the theory of pressure operation this would indeed be true at all frequencies, but here is our first example of how patterns are frequently dependent.

The microphone exhibits what is called the 'obstacle effect'. We might think that a microphone would not disturb the distribution of the energy in a sound wave. But it is found that if any object is placed in the path of a wave its effect on the wave depends on its size relative to the wavelength. If the object is small compared with the wavelength it has no effect on the wave; if the object is large compared with the wavelength it has a great effect on the distribution of the energy in the wave (see Fig. 5.7).

The object shape is important, as well as size, but considering size only at the moment we can see that if the object is a microphone its effect on sound waves will very definitely depend on frequency. At low frequencies—long wavelength—it will have no effect and the directivity pattern will be omni-directional. At high frequencies—short wavelengths—the microphone will cast 'acoustic shadows' because it is larger than the wavelength. If we think of the microphone as facing away from the source, we can see straight away that the diaphragm will be in the acoustic shadow—an area where the sound energy is heavily attenuated. If the microphone is now turned towards the source, the output of the microphone will rise. In other words the directivity pattern is no longer omni-directional but is directional—the microphone is most sensitive when it is 'looking' direct at the source (see Fig. 5.8a).

One might say 'why not reduce the size of the microphone?'. But the output voltage will depend, in the first instance, on how much of the sound wave affects the diaphragm. This must depend on the size —more strictly the area—of the diaphragm, and so it cannot be reduced too far otherwise the sensitivity of the microphone will be too low.

We thus see that the directivity pattern of a pressure operated microphone tends to be omni-directional at low frequencies and gradually becomes increasingly directional as the frequency rises. There is one design of a spherical, pressure operated microphone which has a special attachment to suppress the results of the obstacle effect. This consists of cloth stretched between two woven metal covers. In circumstances where a directional pattern is required, it could be obtained by using a pressure operated microphone which exhibits the obstacle effect and putting a bass-cut filter in the circuit. This of course attenuates all the low frequencies and gives a restricted frequency response, but over this range the microphone has reasonable directivity. It is a trick worth trying if and when the need arises.

Bi-directional. This is often referred to as a 'figure-of-eight' since 8 does describe the pattern fairly accurately. It is obtained by using a pressure gradient operated system, such as the ordinary ribbon microphone.

We saw earlier in the chapter that when there is a difference in path length between the two faces of the ribbon or diaphragm, a pressure difference will exist to make the diaphragm vibrate. When a ribbon microphone is facing the source, the difference in path length, d, is a maximum. If the source now moves round the microphone, keeping a fixed distance from it, d must decrease, becoming zero when the source has moved through 90°—the microphone is sideways on to the source and so the distance from the source to either side of the ribbon is the same. After passing through this zero, d will increase until it reaches the maximum value again, when the source is directly at the back of the microphone—it has moved through a total of 180°. While the source is moving back to its starting position, d will pass through another zero at 270°, and will increase again to its maximum value when the microphone is once more facing the source (Fig. 5.8b).

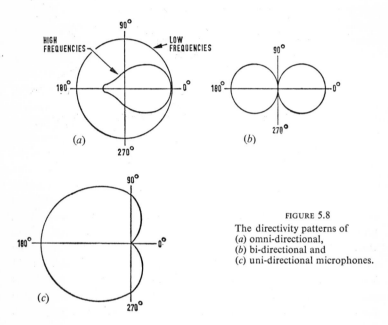

(a)

(b)

(c)

FIGURE 5.8
The directivity patterns of
(a) omni-directional,
(b) bi-directional and
(c) uni-directional microphones.

By similar reasoning, one can see that the vertical pattern must also be a figure of eight. Simple trigonometry shows that the effective path length is $d \cos \theta$ where θ is the angle of incidence.

An important point to note about the pattern is that there is a phase reversal in the output when d passes through a zero. Although the output at 180° is of the same amplitude as at 0°, it is of opposite phase. This has to be considered, for example, when two ribbon microphones are close to each other, side by side. It might happen that one microphone has its 'face' towards the source while the other has its 'back' towards it; the mixed output would be seriously reduced. With two identical microphones, and under certain conditions, the mixed output would be zero. Reversing one microphone, or reversing its leads, will avoid this. Whenever two microphones are being used, although the microphones would not normally be used close together, it is wise to carry out a reversal test to check that phasing is correct (see Fig. 5.9).

By and large, figure-of-eight patterns are much more nearly independent of frequency than are omni-directional ones. In ribbons especially, if the permanent magnet is placed at the bottom of the

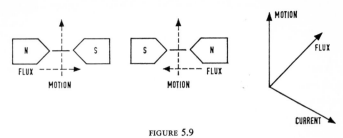

FIGURE 5.9

This shows how two ribbon microphones placed close to each other can produce a cancellation effect when their outputs are mixed. By applying the well-known Right-hand Rule which relates the motion of a conductor, the direction of the magnetic flux and the resultant current, one can see that the flow of currents in the ribbons would be in opposite directions. Turning one of the microphones around *or* reversing the connections to one of them will cause the microphones to operate in phase.

pole-pieces and both sides of the ribbon are open to the sound waves, the patterns obtained are virtually independent of frequency in both the horizontal and vertical planes. The magnet in the case has to be powerful and hence expensive. One way out of this difficulty is to use smaller magnets—say three—and space them evenly across the pole-pieces. This, of course, causes the microphone to be asymmetrical and the directivity pattern is affected, but not seriously (Fig. 5.3).

Uni-directional. 'Cardioid' is the word used to describe a one-sided pattern, and ideally it should be heart-shaped. It provides the same degree of discrimination of direct against reflected sound as a figure-of-eight pattern. The high ratio of front to back pick-up is very useful in practice (see Fig. 5.8c).

The pattern could *in theory* be obtained by combining the electrical outputs of a pressure operated microphone and a figure-of-eight one. Remembering that the omni-directional pattern does not alter its phase, whatever the angle of incidence of the sound wave, whereas the bi-directional pattern does change its phase, we can see that adding them will provide a uni-directional pattern (see Fig. 5.10).

When both microphones are facing a source, if their outputs are then in the same phase, the outputs will be of opposite phase for sounds arriving from the rear. And if they are equal in amplitude,

92

the two signals will completely cancel, giving a combined output of zero. For sounds approaching from the sides, only the omni-directional microphone will be effective and the output will thus only be one-half that of the two units in phase.

At the beginning of an earlier paragraph it was said that 'in theory' the uni-directional pattern could be obtained by combining a pressure operated unit and a pressure gradient operated unit. Practical difficulties are immediately obvious; two microphone units have to be fitted inside the same case and this is therefore likely to be bulky; the frequency responses of the two units would have to be identical; the directivity patterns of the two units should be consistent at all frequencies. But we know that the third requirement is most difficult to meet—so-called omni-directional microphones tend to possess a circle directivity pattern at low frequencies only. Various practical tricks are used by manufacturers to overcome these difficulties, but the method of combining two such units is not widely used.

It would obviously be a great advantage if we could obtain the uni-directional pattern by using only one microphone element. Single diaphragm cardioids have indeed been developed and are now widely used. They can be moving coil, ribbon or condenser types and this shows that the particular method of generating the electrical output is immaterial. This is because the pattern is derived by combining

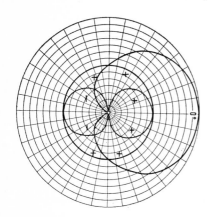

FIGURE 5.10
Because the bi-directional microphone changes the 'sense' of its output for sounds approaching the rear face, and the phase of an omni-directional microphone remains the same irrespective of the angle of incidence, the two responses can be combined to provide a cardioid pattern.

the driving forces on the diaphragm due to a pressure component and a pressure-gradient component. The combination is performed at the acoustic input, not on the electrical output as in the previous method (see Fig. 5.11).

With the single diaphragm cardioid microphone, the sound wave has access to both sides of the diaphragm, but the rear face is not

93

freely exposed. It is enclosed by an acoustic chamber. Entry to this chamber is by holes or slots in the microphone casing. The chamber controls the phase of the sound striking the rear face of the diaphragm relative to that striking the front face. In fact this method was originally called the 'phase shift' principle.

The acoustic distance between the two faces of the diaphragm is important, and Perspex discs are fitted on some cardioid microphones to obtain the correct distance for the particular microphone.

Variable-Pattern Microphones

In professional work it is often a great advantage to be able to vary the directivity pattern of a microphone. It is helpful in controlling the balance in a studio on some occasions, without having to move the performers or microphones.

One method used was to have a ribbon microphone with a shutter which could be pulled across to form an air chamber at the back of the ribbon. Obviously the pattern obtained depended on how much of the ribbon was obscured. When the shutter was fully over, the microphone behaved like a pressure operated system and gave an omni-directional pattern. When the ribbon was freely exposed, the shutter out of the way, a figure-of-eight pattern resulted. Between these two extremes, intermediate patterns could be obtained by suitable positioning of the shutter.

FIGURE 5.11

Single-diaphragm cardioid operation. For convenience we split the total path length into two components: '*de*' the external part and '*di*' the internal part.
Sound approaching from 0° must traverse '*de*' and then '*di*'. For sounds approaching from 90° or 270° '*de*' is ineffective; only '*di*' will produce a difference in pressure. For sounds approaching from 180° the two path lengths will be traversed simultaneously. If '*de*' and '*di*' can be made equal, the driving force—and hence the microphone output—at 90° and 270° will be half that at 0°. For sounds at 180° the driving force will be zero since sound will arrive at the back of the diaphragm at exactly the same time as sound reaches the front since the path lengths are the same.

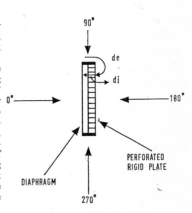

Another method is to use two cardioid units placed back-to-back so that their directions of maximum sensitivity are 180° apart. For example, condenser microphones using this technique are available where the pattern is controlled by altering the voltage on one of the diaphragms relative to the common fixed plate. This causes the two

94

units to combine in phase or out of phase, the pattern varying from omni-directional, through cardioid to figure-of-eight (see Fig. 5.12).

Sensitivity

The electrical output of a microphone for a given sound strength should of course be as high as possible, so that an adequate signal-to-noise ratio is obtained at the start of the electrical chain from microphone to recorder or from microphone to loudspeaker.

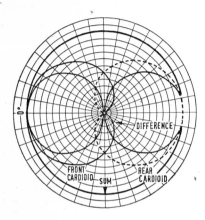

FIGURE 5.12
This shows how an omni-directional pattern and a 'figure-of-eight' pattern can be obtained by adding and subtracting the outputs of two cardioid units.

Since the microphone converts acoustical signals into electrical signals, the definition of sensitivity should give the strength of the electrical signal output for a given acoustic input. Unfortunately there are several methods used by manufacturers in giving a microphone's sensitivity.

One method is to measure the voltage output when an acoustic pressure of 1 dyne/sq. cm is actuating the diaphragm. This voltage can then be referred in decibels to an agreed reference voltage—normally taken to be 1 V. Thus the output of the microphone is stated to be 'x dB relative to 1 V per dyne per sq. cm'. To avoid any possible effect which the following equipment, such as an amplifier, might have on the output voltage, no equipment is connected to the microphone; i.e. it is on 'open-circuit'.

One objection to the above method is that a pressure of 1 dyne/sq. cm is impractically low, and a more representative value of 10 dyne/sq. cm is sometimes used. This raises the stated output voltage by 10 dB, once again the microphone being on open-circuit.

Another method, given in a Radio and Television Manufacturers' Association specification refers the power output of the microphone in dB to the agreed zero power level of 1 mW for an acoustic pressure

95

of 0·0002 dynes/sq. cm. This, as we saw in Chapter 2, is the auditory zero level of acoustic pressure at a frequency of 1 kc/s.

Impedance

Obviously the output impedance of a microphone must be known before it is connected to say an amplifier or tape recorder. In practice, the impedance of microphones can range from a few ohms to several thousand ohms.

Moving coil and ribbon microphones are low impedance generators, typical figures being about 30 ohms and 1 ohm, respectively.

With the ribbon, a transformer is normally built inside the case to step up the impedance, the turns ratio of the transformer controlling the output impedance. Some models have fixed impedances, typical values being 30 ohms or 300 ohms. With other types, a choice of impedance is available and this can be Low—25 ohms, Line—600 ohms or High—50,000 ohms.

The advantage of having a low impedance microphone is that the length of cable connecting it to the rest of the equipment can be as long as one wishes, 100 yd or so. The disadvantage is that a transformer must be used, if one has equipment with a high input impedance, to match the microphone and its cable into the equipment. In addition, the sensitivity of low impedance microphones is generally low.

A high impedance microphone can be coupled straight into the equipment without a transformer, and the sensitivity is usually much higher, 30 dB greater is typical. But the length of cable which can be used with a high impedance microphone is severely limited. This is because the cable acts as a capacitance across the microphone output terminals. With a low output impedance, the effect of this shunting is negligible, but with a high output impedance the shunting causes a serious drop in the signal passed on to the apparatus. The value of the capacitance due to the cable depends on its length, and cables can be specified as having so many units of capacitance per foot. The unit is the picofarad, normally written as pF. It is a millionth millionth part of the basic unit, the Farad (i.e. $1 \text{ pF} = 1 \text{ F} \times 10^{-12}$). One manufacturer quotes 80 pF as being the maximum value of shunting capacitance that can be tolerated before serious attenuation will result. Depending on the type of cable, up to 20 feet is about the maximum length that can be used with safety.

Moving coil microphones are available without transformers giving an output impedance of 30 ohms. However, like the ribbon, there are many models available with a built-in transformer to give a choice of output impedances, and the foregoing remarks apply equally well to the moving coil types.

Crystal microphones have a very high output impedance. They consist essentially of two metallic skins separated by the crystal material, and so they behave rather like a capacitor—in practice a value of about 1,000 pF is representative. This means that not only is the impedance high, but it is also frequency dependent since capacitive reactance is inversely proportional to frequency. With decreasing frequency the internal impedance of the microphone goes up, and this can lead to a loss of bass frequencies unless one feeds the microphone into a much larger impedance. This is illustrated in Table 5.1.

TABLE 5.1. *Matching of High Impedance Microphone*

Input Impedance of Equipment (Megohms)	Frequency at which response is 3 dB down on that at 1000 c/s (cycles per second)
1	200
2	100
5	40
10	20

With such high impedances, short cables of the order of 8 feet, are essential. Additional cable length adds shunt capacitance which will reduce the output. When the lead capacitance equals the capacitance of the microphone, for example, the reduction is 6 dB.

All high impedance, low capacity cables are susceptible to hum and interference pick up and can generate noise signals if they are moved. It is therefore essential to take care with the cables of high impedance microphones.

With some moving coil ribbon microphones, the output impedance rises at low frequencies. One example is quoted as having an impedance of 25 ohms at mid-frequencies and 35 ohms between 70 and 100 c/s. If the microphone were matched at the mid-band value, there would be attenuation of the bass frequencies and manufacturers usually advise operating the microphone into an impedance much greater than the nominal output impedance. In the case quoted, the impedance should be above 200 ohms. Another manufacturer produces a moving coil microphone which has an output impedance of 200 ohms at the mid-frequencies, and recommends that it should be worked into an impedance of not less than 600 ohms.

Condenser capsules too have a high impedance, but it is much

G

higher than a crystal type, the capacitance being about a tenth that of a crystal element. This often means in practice that an amplifier is placed immediately adjacent to the capsule so that the leads are kept as short as possible. In fact a professional condenser 'microphone' normally consists of the capsule with its associated 'head' amplifier. The performance of the microphone is the performance of the two together. This, of course, makes condenser microphones expensive, and they are rarely used in the amateur field.

REVIEW OF MICROPHONE TYPES

In this section, several microphones will be examined to see how the basic principles outlined above have been employed by the microphone designers.

Pressure Operated Microphones

Moving Coil

Standard Telephone and Cables Microphone 4021. This is an example in which the designer has used two ideas to help reduce the obstacle effect of the microphone. The first feature to notice is that the case is spherical. This shape gives least disturbance to a sound wave for a given volume. The second feature is the

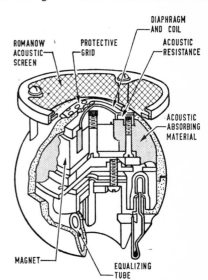

FIGURE 5.13

Cut-away model of the S. T. & C. 4021. By using a spherical case and an acoustic screen in front of the diaphragm the variations in high frequency performance with different angles of incidence can be smoothed out.

porous front screen. This attenuates the high frequency sounds arriving on the axis and, by reflection, enhances high frequency sounds arriving from below.

The response curves show how the microphone's frequency and directivity responses are maintained over a wide range.

It was mentioned earlier (see p. 90) that one could increase directivity at high frequencies by fitting a solid baffle around a microphone. A baffle is available for the 4021, and when this is fitted in place of the usual screen, it gives the response curves shown in Fig. 5.14, which illustrate how the high frequency response is altered.

The cut-away illustration (Fig. 5.13) shows how the inside of the microphone is arranged. The diaphragm is made of 0·0004-in.-thick aluminium alloy and the

TYPICAL FREE FIELD FREQUENCY RESPONSE CURVE($0dB = IV/DYNE/cm^2$—OPEN CIRCUIT)

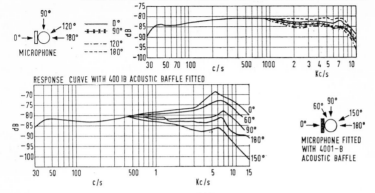

FIGURE 5.14

For some purposes, P. A. work is an example, a rising frequency
response is sometimes useful. By replacing the acoustic screen on
the S. T. & C. 4021 microphone with an acoustic baffle this can be
obtained, as can be seen in the lower diagram.

coil is wound from aluminium tape. The diaphragm is domed and corrugated to
make it move like a piston and not break up into various resonant modes of
vibration. The mass of the moving coil system is 80 milligrammes and the output
voltage is approximately 100 µV or 80 dB relative to 1 V/dyne/sq. cm. The absorbent material is included to prevent acoustical standing wave effects. The
equalising tube is used, coupled with the air inside the case, to improve the low
frequency response as in a vented loudspeaker cabinet.

This microphone is essentially an indoor instrument and should not be used
out-of-doors in a severe wind.

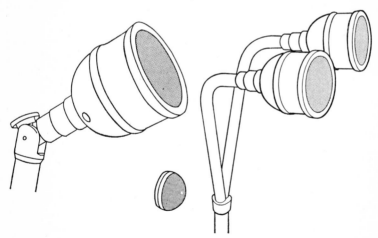

FIGURE 5.15

The S. T. & C. 4035 moving coil microphone. The small inset
(centre) shows the windshield designed for this microphone, and a
twin mounting is shown (right).

99

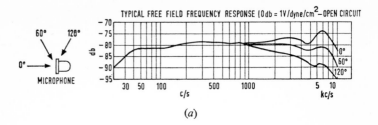

(a)

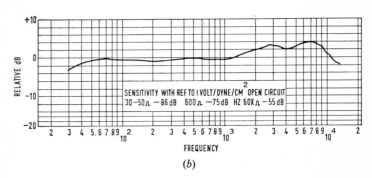

(b)

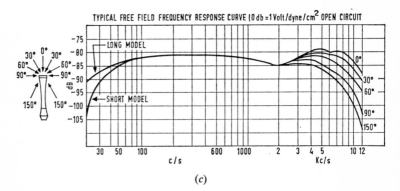

(c)

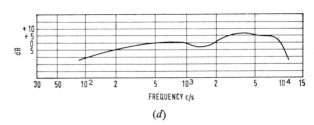

(d)

FIGURE 5.16
Response curves for (a) S. T. & C. 4035, (b) Tannoy 'Slendalyne',
(c) S. T. & C. 4037, (d) A.K.G. D7A.

100

Standard Telephone and Cables Microphone 4035. This microphone is intended for use out-of-doors, and has been designed with a near spherical shape to minimise wind noise. It is fitted with protective screens to avoid the ingress of rain. When used with the windshield, which consists of rubberised hair covered with woven stainless steel, there is an effective reduction in wind noise of some 15–20 dB (see Fig. 5.15).

The microphone should face the sound source and it is not omni-directional at high frequencies. The response curves show how the output falls off above 1,000 c/s (see Fig. 5.16a).

The output voltage is 128 μV or 78 dB relative to 1 V/dyne/sq. cm.

Tannoy Slendalyne. This microphone is of small size and has a diaphragm assembly which is moulded in one piece from a stable polyester film. Its small size maintains its omni-directional response up to 5,000 c/s approximately. A

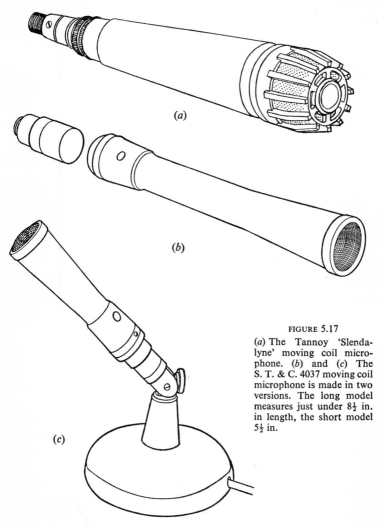

(a)

(b)

(c)

FIGURE 5.17

(a) The Tannoy 'Slenda-lyne' moving coil micro-phone. (b) and (c) The S. T. & C. 4037 moving coil microphone is made in two versions. The long model measures just under 8½ in. in length, the short model 5½ in.

built-in transformer allows for the selection of convenient impedances. In the low impedance case, the sensitivity is —86 dB and at high impedance (60 kΩ) it is —55 dB (both figures relative to 1 V/dyne/sq. cm).

Standard Telephone and Cables Microphone 4037. This microphone is designed to be of light weight and unobtrusive appearance. The photograph shows the A version with its tapered tubular case. There is a B version which is shorter in length. The response curves show how the high frequency response is affected above 3,000 c/s by the angle of incidence of the sound waves. It can be used successfully out-of-doors, in severe wind conditions, and a plastic foam windshield can be fitted. The sensitivity is —84 dB relative to 1 V/dyne/sq. cm.

A.K.G. D.9. This is a small desk microphone which has been produced mainly for the recording of speech on tape recorders. The frequency range is 80–10,000 c/s and the manufacturers have given it a rising response since this increase in the upper frequencies can, in many cases, improve intelligibility. The directivity pattern is omni-directional. Two impedances are available—200 ohms, in which case the sensitivity is —74 dB relative to 1 V/dyne/sq.cm, and 50,000 ohms, which gives a sensitivity of —52 dB.

R.C.A. BK-6B. This small, lightweight microphone can be attached to the person, and is widely used in professional work. Being under three inches long, about one inch in diameter, it weighs only 2⅓ oz, and this makes it easy to fix on the chest of the speaker. A lanyard is supplied to secure it in position.

In Chapter 4 we saw how the voice is projected from the mouth with the distribution becoming very directional at the important high frequencies. A microphone on the chest will obviously tend to miss these frequencies, and in addition it will pick up the low pitched chest noises which we produce when speaking. To overcome these potential defects due to the microphone position, the frequency response and directional response have been 'tailored'. The bass frequencies are attenuated to cut down the chest sounds, and the high frequency response is increased to restore the crispness and definition.

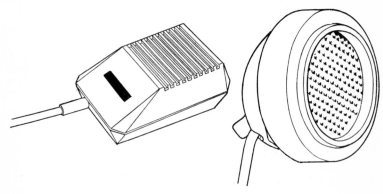

FIGURE 5.18

The A.K.G. D7A (left) is an example of an inexpensive pressure operated moving coil microphone. The ACOS Mic. 55 (right) is designed to fit on to the lapel.

The microphone can also be used in the hand or in a holder. In these circumstances, the general rule is to talk across the instrument, but with some speakers it is worth experimenting with them talking directly into the microphone, the position adopted depending on the response required for the particular speaker's voice.

This is another microphone using a plastic diaphragm. It is of low impedance so that a long cable can be used, there being a choice between 30, 150 or 500 ohm. The sensitivity is —85 dB relative to 1 V/dyne/sq. cm.

ACOS. Mic. 39 (Dynamic). This is a small microphone giving an omnidirectional pattern. The frequency range is within 3 dB from 80 to 10,000 c/s. and

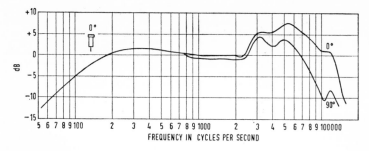

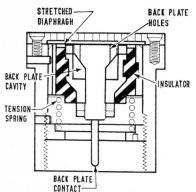

STRETCHED DIAPHRAGH

BACK PLATE HOLES

BACK PLATE CAVITY

INSULATOR

TENSION SPRING

BACK PLATE CONTACT

FIGURE 5.19

The upper diagram shows how the response of the R.C.A. BK-6B moving coil microphone, designed to be worn on the chest, has been adjusted to offset the frequency discrimination of the unusual working position. The sectional view (left) is of an omni-directional condenser microphone used in acoustic standardisation work. (Courtesy S. T. & C. Ltd.)

is 10 dB down at 50 c/s and 15,000 c/s. Two impedances are available 200 Ω and 50 kΩ, the sensitivities being —80 dB and —54 dB relative to 1 V/dyne/sq. cm.

Crystal

ACOS. Mic. 39 (Crystal). This microphone is available in a moving coil version. The attractive cheapness of the crystal element can be appreciated when one sees that it costs less than half the price of a moving coil one. Of course the quality of the moving coil in terms of frequency response is better. It is a slim, lightweight instrument suitable for tape recording and gives approximately 1 mV output (—62 dB relative to 1 V/dyne/sq. cm).

The capacity of the element is 880 pF and the load resistance should be 5 mΩ. The recommended lead length is 8 feet.

ACOS. Mic. 55. On many occasions it is very useful to be able to have both hands free when using a microphone. This small microphone is designed to fit on to the lapel. It has a relatively high sensitivity, —58 dB relative to 1 V/dyne/sq. cm. The directivity pattern is omni-directional but of course this is much modified when it is worn.

Pressure Gradient Microphones

Standard Telephone and Cables Microphone 4038. This is a high grade microphone designed especially for studio work and it contains several interesting features. The ribbon is a strip of aluminium leaf, 1 in. long and ¼ in. wide. The moving mass of this is about 0·2 mg compared with about 80 mg for the diaphragm in moving coil microphones. The large permanent magnet is mounted below the pole pieces and this avoids the bulk of the magnet casting any acoustic shadows. With a short ribbon, there is little loss due to phase difference along the ribbon for sounds approaching in the vertical plane and the bi-directional response is well maintained at all frequencies.

103

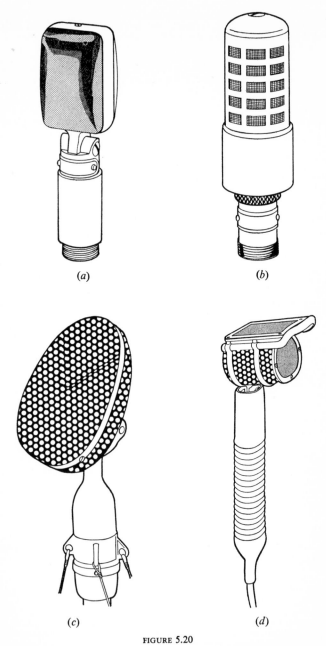

(a)   (b)

(c)   (d)

FIGURE 5.20

Four types of ribbon microphone. (a) Reslosound Type RB, (b) Reslosound Type PR, (c) S. T. & C. 4038, (d) S. T. & C. 4104 noise-cancelling type.

Since the ribbon is tensioned it behaves like a violin string and would produce resonant peaks unless it were damped. This is done by trapping the air surrounding the ribbon between screens of fine woven metal close to each side of the ribbon. This damping action will, of course, reduce the low frequency output and to counter-balance this a fine silk gauze 'halo' is fitted round the pole-pieces. The gauze increases the path length between the faces of the ribbon and this increases the driving force. Since the driving force rises with frequency in a pressure gradient operated microphone, the effect is to produce a larger proportional change at low frequencies than at high. The bass frequencies are therefore boosted.

Ideally, the outer case of a microphone should not introduce any acoustic effects due to reflection. In practice, however, the casing usually causes standing

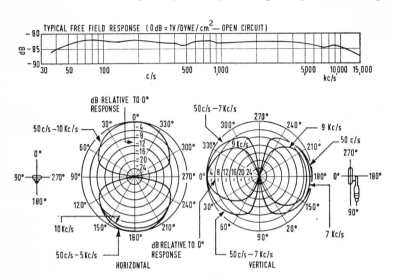

FIGURE 5.21

The frequency response and directivity patterns of the S. T. & C. 4038 high quality ribbon microphone.

wave effects due to reflection between it and the pole-pieces. These cause the output to rise at one frequency and dip at a higher one. By suitably adjusting the opening between the casing at the ribbon and pole-pieces, these frequencies can be chosen so that the rise occurs at a frequency where the output is falling off, and the dip at a frequency outside the normal range of hearing. In the 4038 this is done by 'dishing' the case, so that it is brought closer to the pole-pieces and ribbon.

A ribbon microphone such as the 4038 cannot be used out-of-doors. Even in very low wind velocities the response would be marred by heavy rumbling. In high wind velocities the microphone could be damaged. The microphone is usually used stationary, otherwise it will produce rumble. It is a low impedance instrument of 30 ohms and has a sensitivity of —85 dB relative to 1 V/dyne/sq. cm.

Reslo Ribbon Microphones. This Company offers several types of ribbon microphones. They have a different magnet arrangement than in the S. T. & C. 4038. A lighter and cheaper microphone can be produced by connecting small magnets directly across the back of the pole-pieces. There are usually three, one at each end and one in the middle. This makes the magnet system more efficient for a given volume of permanent magnet and it allows the use of smaller magnets— hence they can be cheaper.

Of course this method makes the microphone asymmetrical acoustically, the directivity pattern is not bi-directional at all frequencies and the frequency

105

response is not so extended. But the microphone is much lighter; 10 oz would be a representative figure whereas the S. T. & C. 4038 weighs about 2½ lb. The manufacturers use this asymmetrical system in some of their models to protect the ribbon when the speaker, or singer, is close to the microphone. If a more distant technique is used, say over two feet, the unit can be turned right round to allow the sound direct access to the ribbon. This gives a better frequency response.

Type RB. The case in this microphone is made of two shells. It is assembled with the ribbon unit towards the rear shell and the magnet assembly towards the front one. Although this protects the ribbon as outlined above when a close technique is used, the manufacturers provide a felt pad which can be fitted inside the front shell to prevent possible damage due to condensation of the performer's breath causing corrosion and rusting of the ribbon and magnet assembly. They also supply further pads which give some control over the directivity pattern and frequency response. These are made of felt and fabric and, by suitable combination of these, the directivity pattern can be made to have considerable discrimination against sounds coming to the rear of the microphone. A reduction in bass response is also possible.

The matching transformer is in the tubular mounting which also contains the connecting socket. Various impedances are possible ranging from 30 up to 10,000 ohms. The sensitivity is —81 dB relative to 1 V/dyne/sq. cm.

Type VRT. This is a 'broadcasting' version of the RB type. The ribbon element is mounted in the front shell and the magnet assembly in the rear one. This arrangement, as was explained above, gives a better frequency response which is further improved by having a higher quality transformer to match the ribbon to the microphone cable. Two impedances are available, one 30–50 ohms and the other 300 ohms. In the latter case, the sensitivity is —81 dB relative to 1 V/dyne/sq. cm.

Type PR. By simplifying the construction, the manufacturers have been able to produce a cheaper version of the RB microphone which is suitable for domestic recording work. The magnetic element has been miniaturised to allow a 'Pencil' form of assembly. As with the other microphones, the PR is used with the ribbon facing the performer if the working distance is at least a foot. For closer use, the microphone is turned round so that the magnets give some protection to the ribbons.

To control the directivity pattern, plastic foam pads are fitted behind the ribbon in between the magnets. These pads also act as damping to suppress the main ribbon resonance. They behave as phase shifting devices for sound striking the rear face of the ribbon. As we saw earlier, this will lead to a cardioid type of directivity pattern. However, this method will not function properly over the whole audio frequency range, and in fact the roughly cardioid pattern is only produced at high frequencies. At other frequencies the basic bi-directional pattern is produced. The plastic pads can be removed if a pattern approximately bi-directional at all frequencies is required.

The impedances available range from 30/50 ohms to 10,000 ohms and the sensitivity at the low impedance is —86 dB relative to 1 V/dyne/sq. cm.

Standard Telephones and Cables Ltd.: Microphone 4104. Although this microphone is normally only used in professional work, it is felt that it might be of interest to amateur readers.

In discussing ordinary ribbon microphones, we saw how the ribbon has to be protected if a very close working distance is used. Provided this is done, the increase in output at low frequencies can be exploited to give good reproduction of speech from noisy surroundings. There is a great deal of low and middle frequency energy in such noise and, if it can be attenuated, good discrimination will be achieved. A fuller explanation will be given in the next chapter but a simple description at this stage will be helpful. This energy will approach the microphone as plane waves, whereas the energy from the close voice will be spherical. By acoustical filters, the frequency response for the spherical waves is made substantially flat and this means that the response for the plane waves is very much reduced. Of course the speaking distance must be fixed, so that the response of the filters is matched to the increase in low frequency output. This is done by having a positioning bar which comes in contact with the speaker's upper lip and fixes the distance at about two inches.

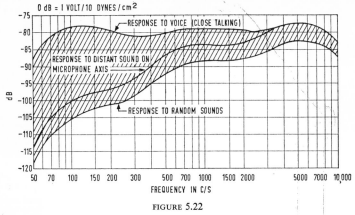

FIGURE 5.22

The frequency response graphs show how effective a noise cancelling microphone can be. This data is for the S. T. & C. 4104 ribbon microphone.

At 50 c/s the microphone gives 30 dB discrimination between the close source and background noise of random incidence; at 500 c/s it is 15 dB and at 2,000 c/s it is about 10 dB.

The design of the microphone has also to take into account the difficulties of getting good speech quality with such a close working distance. During speech, there are four sources of sound, the mouth, the nose, the throat and the chest. The last two are of secondary importance and may be ignored in this discussion. With the mouth and nose, there are difficulties due to breath noises and the effects of explosive speech sounds. To illustrate the latter point, mention can be made of the great increase in air particle velocity when one is enunciating certain consonants.

To get natural sounding speech, a balance must be obtained between the mouth and nose sources. And the question of frequency response must be considered since subjective experiments indicate that a slight rise in the high frequency response is desirable between 4 and 7 kc/s to offset the different frequency spectrum in close range speech. To control breath noises, stainless steel woven meshes are used to act as shields, one at the front of the microphone for the mouth, the other on top so that, when the microphone is in position, it comes just under the nose.

Although ribbon microphones designed for studio work cannot be used out-of-doors because of wind causing excessive rumbling in the output and possible damage to the ribbon, the lip ribbon can be used in wind velocities up to 20 m.p.h. and it will not be unduly affected. Simple wind shielding doubles this figure and still gives acceptable results.

Since the microphone is likely to be held for long periods, weight is an obvious problem. With the 4104 this has been reduced to 10 oz.

The microphone has an impedance of 30 ohms and its sensitivity is —82 dB 10 dyne/sq. cm² relative to 1 mW. Note that the reference level used in specifying this microphone takes into account the special way it is used.

Cardioid Microphones

We have seen how the heart-shaped directivity pattern gives a uniform response over a solid angle of about 120° and discriminates against sounds coming from outside this area. This response makes the cardioid very useful in many applica-

107

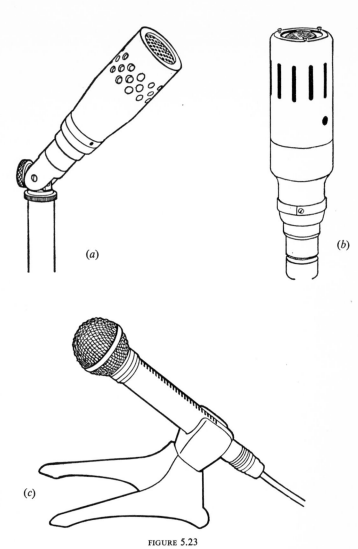

FIGURE 5.23

The apertures leading to the acoustic chambers behind the diaphragm can be clearly seen in these three types of moving coil cardioid microphone; (*a*) S. T. & C. 4105, (*b*) S. T. & C. 4106, (*c*) A.K.G. D.24.

tions; in public address systems, for example, where it can help to avoid howl-round due to feedback between the loudspeakers and microphone; in theatres, to reduce the effect of audience noises, and in studios where it can again cut out sounds from unwanted sources and also give a good ratio of direct to reflected sound, which is useful when the acoustics tend to be too live.

Combination cardioids were at one time the only way in which the more rugged electro-dynamic units—moving coil and ribbon—could be used. Single diaphragm condenser microphones have been in use for many years, but their good performance both in terms of frequency response and constancy of directivity pattern is reflected in their high cost and the need to handle them very carefully, which limits their use to professional work. Nowadays, single-diaphragm electrodynamic units are available either using a ribbon or a moving coil. These are obviously much lighter and smaller than the combination ones. One manufacturer offering both types quotes the weight and dimensions as follows:

Single Diaphragm Moving Coil. 6·3 oz: 3·6 in. long and 1·47 in. diameter
Combination Moving Coil and Ribbon. 3 lb: 9 in. × 3 in. × 3½ in.

The phase-shift principle used in single-diaphragm units has been previously explained (see p. 93). Basically it consists of a series of holes or slots working into various cavities inside the microphone. The actual way in which the principle is applied depends, of course, on the manufacturer but a basic problem is that it is difficult to maintain a good cardioid pattern over a wide frequency range.

With moving coil and ribbon units, difficulties arise because the mechanical resonance occurs at a low frequency. We saw in the case of the ribbon micro-

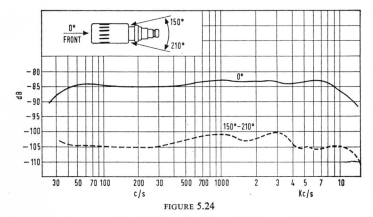

FIGURE 5.24

Frequency response on axis and at the angles shown of the S. T. and C. 4106 cardioid.

phone how damping screens had to be brought close to the ribbon because it was resonating at about 40 c/s. It was essential for the microphone to have a mechanical resonance below the lowest audio frequency we are interested in, since this corrects the rise in driving force with frequency due to the pressure gradient principle.

This can be likened to the way in which a moving coil loudspeaker works. It is well known that, for good low frequency response, the speaker should have a very low resonant frequency if the radiated output is to be independent of frequency. Now it is comparatively easy to provide a very low resonant frequency with a ribbon, but with moving coil units it is difficult to achieve since the coil must be positioned accurately in the centre of the narrow magnetic gap. This is done by having sufficient stiffness at the rim of the diaphragm, but the result is that the resonance occurs at too high value for the low frequency response to be maintained. Reducing the diaphragm rim stiffness would obviously reduce this, but it would leave the coil so loose that it would be prone to shock—as is the case with the ribbon microphone.

This problem is tackled by coupling to the back of the diaphragm acoustic chambers which reduce the diaphragm resonance to a low frequency. In one case, the freely vibrating diaphragm with the moving coil has a resonance of 300 c/s. When the microphone is assembled, the acoustic chambers reduce this to about

109

90 c/s. How these chambers are laid out and arranged of course depends on the designer.

At high frequencies, difficulties arise because of the possibility of reflections inside the unit causing standing waves and of the unit causing an obstacle effect and upsetting the distribution of the sound energy. This is where the condenser method scores; the capsule is small and the mechanical resonance occurs at a much higher frequency.

Since the driving force on the diaphragm must contain a pressure gradient component, a close working distance causes the low frequency output to rise, though not so seriously as when pressure gradient working is used on its own, as in the ribbon microphone. Some manufacturers, usually Continental ones, provide a switch on the microphone to give some bass cut, around 8–10 dB at 50 c/s is a representative figure, when the speaker is close. In some moving coil units, especially speech reinforcement microphones, the frequency response is allowed to fall off at the low frequencies, say below 500 c/s, and this is compensated by the rise when the speaker uses a working distance of about twelve inches.

Standard Telephone and Cables Ltd.: Microphone 4105. This is a moving-coil unit designed principally for sound reinforcement systems. Holes in the side of the case act as inlets to the rear face of the diaphragm and form part of the phase-shifting network. The diaphragm, made of plastic, can withstand considerable temperature and mechanical changes. The frequency response falls by 10 dB at 50 c/s compared with that at 1,000 c/s. The discrimination is better than 15 dB between the front and rear responses. The sensitivity is —82 dB relative to 1 V/dyne/sq. cm. The output impedance is normally 30 ohms, but the input impedance of the amplifier into which it is fed should be not less than 200 ohms.

Standard Telephone and Cables Ltd.: Microphone 4106. This moving coil cardioid is produced for high quality studio work and has a frequency response within 3 dB from 40 to 10,000 c/s. It is based on the same design principles as the 4105 and again uses a plastic diaphragm. The discrimination between front and rear responses is 20 dB. The sensitivity is —85 dB relative to 1 V/dyne/sq. cm and the impedance is 30 ohms at the middle frequencies. As with the 4105, the microphone should be worked into an impedance of not less than 200 ohms.

A.K.G. Microphone D.19. This is an interesting example of a moving coil cardioid since it uses a slotted tube at the back of the diaphragm as part of its

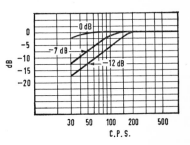

FIGURE 5.25

With some types of moving coil directional microphone it is possible to attenuate the bass response when recording speech at close range. The graph shows the degrees of attenuation available in the A.K.G. D.25 Moving Coil Cardioid.

phase shifting network. The tube increases the driving force of the pressure gradient component and also decreases the resonant frequency of the diaphragm. Special precautions ensure that the tube does not resonate or produce standing wave effects.

The microphone is primarily designed for high quality domestic tape recorders and has a quoted frequency range of from 40 to 16,000 c/s. A bass switch introduces 10 dB attenuation at 50 c/s for close working distances. The discrimination between front and rear responses is about 15 dB.

There is a choice of output impedances between 200 ohms and 50,000 ohms and the sensitivity is —75 dB and —52 dB respectively (both referred to 1 V/dyne/sq. cm).

A higher quality version of the microphone is marketed with the number D.24.

The frequency response is extended at the bass end and the discrimination between front and rear response is increased to about 20 dB. It has an output impedance of 200 ohms and a sensitivity of —75 dB relative to 1 V/dyne/sq. cm.

Condenser Cardioids

As has been said previously, condenser microphones produce the best cardioid response. Their small size, with diameters of the order of 1 in., and the accuracy with which the response can be adjusted, give professional standards to their performance.

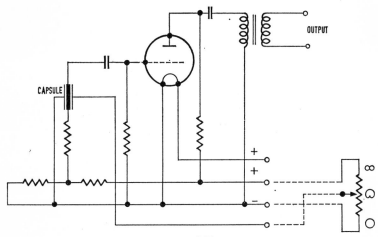

FIGURE 5.26

The circuit arrangements of the A.K.G. C.12 microphone to provide the variable directivity pattern characteristics. The rear diaphragm is connected to the variable resistance in the remote control box. The rigid plate and the front diaphragm potentials are fixed. The other components in the circuit are used in an arrangement typical of condenser microphone head amplifiers. The directly heated valve feeds the output transformer which is connected to a pair of conductors in the multi-core cable.

There are several pressure-operated, condenser microphones available which give an omni-directional pattern but most condenser microphones are used for their cardioid characteristics. With some types the omni-directional pattern is produced by using two cardioid capsules back-to-back, and connecting the two diaphragms together so that they are polarised in the same sense with respect to the central fixed plate. If one wishes a cardioid response, the rear diaphragm is connected to the fixed plate so that, even though it vibrates, there is no contribution to the output since no capacitance exists between it and the fixed plate—the cardioid pattern being produced by the front diaphragm and the fixed plate.

As with moving coil cardioids, adjustment of diaphragm resonance and matching of the pressure and pressure gradient components is essential. To illustrate how precise the manufacture is, one can quote a particular microphone where washers, having a thickness of a few microns (thousandths of a millimetre), are used to control the frictional resistance between the two capsules. Adjustment of this resistance controls the matching of the pressure gradient and pressure components.

Standard Telephone and Cables Ltd.: Microphone 4108. This consists of a small cardioid capsule, approximately 0·875 in. in diameter, which is plugged and screwed on to a head amplifier in a housing of equal diameter. The amplifier is

111

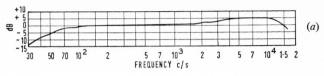

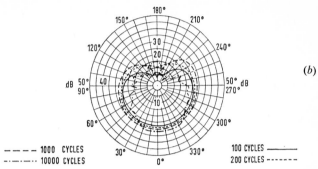

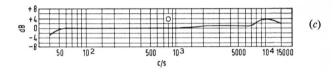

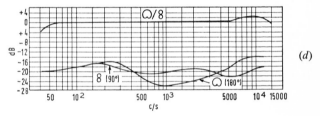

FIGURE 5.27

(a) and (b) The frequency response and polar characteristics of the A.K.G. D24 cardioid.

(c) and (d) The frequency response of the A.K.G. C.12 for three different directivity conditions.

(e) The A.K.G. C.12 pattern selection box.

powered from a separate unit to which it is connected by a special cable which may be up to 100 ft in length.

Uniform directivity of about 25 dB front-to-rear ratio is obtained over a frequency range of from 30 to 15,000 c/s.

The sensitivity is —60 dB relative to 1 V/dyne/sq. cm.

A.K.G. Microphone C.28. This is an interesting microphone since it can be used in three different forms. Basically it is a cardioid unit, but the capsule can easily be unscrewed and an extension tube interposed between it and the head amplifier, the tube acting as a high grade connecting link. The tube is available in two lengths, 18 in. and 46 in. The advantage of this arrangement is that the microphone can be made unobtrusive where this is required. Television Broadcasting, the stage and film studios often require an unobtrusive high quality microphone.

The basic unit has a quoted frequency of from 30 to 20,000 c/s. The sensitivity

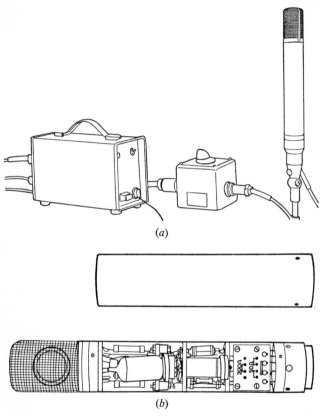

(a)

(b)

FIGURE 5.28

(a) The A.K.G. C.12 condenser microphone with its associated power unit and pattern selectivity box.

(b) The cover removed to show the built-in amplifier.

is —58 dB relative to 1 V/dyne/sq. cm and the output impedance is either 50 or 200 ohms.

A.K.G. Microphone C.12. This uses two cardioid units back-to-back and gives a choice of nine different directivity patterns.

Mention has been made previously of the way in which the potential of the rear

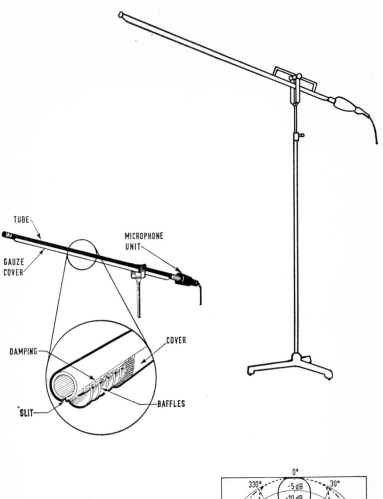

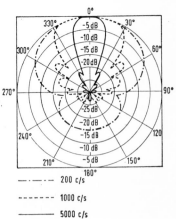

FIGURE 5.29

The Sennheiser Microphone MD82 is a current example of a 'line' type of directional microphone. The graph shows how the directional characteristics vary with frequency. The enlarged view shows the damped slit along the sound collecting tube and the resonant cavities which aid the top response.

—·—·— 200 c/s

– – – – – 1000 c/s

———— 5000 c/s

114

diaphragm can control the directivity pattern (see p. 94). An omni-directional pattern is obtained by connecting the rear diaphragm to the front diaphragm, and a cardioid by connecting it to the fixed plate. In the C.12, the rear diaphragm is connected to a nine position potentiometer. This allows the rear diaphragm potential to be set at the same value as the front diaphragm to produce the omni, or at the same value as the fixed plate to produce the cardioid, or at the same value but opposite sign as the front diaphragm to produce the figure-of-eight pattern.

The changes in pattern do not produce clicks or changes in sensitivity, and the change is a gradual one. It takes a few seconds for the chosen pattern to come into action. The potentiometer is fitted into a separate box, with the directivity patterns marked on the scale, and it can be used at a considerable distance from the microphone. This means that pattern selection can be carried out remotely, and some broadcasting authorities fit the selector box in the control cubicle on the sound control desk. This gives another useful tool in balancing the ratio of direct to indirect sound and in getting separation between the various performers.

The quoted frequency response is 30–20,000 c/s, the sensitivity is —60 dB relative to 1 V/dyne/sq. cm and the impedance can be either 50 or 200 ohms.

Highly Directional Microphones

There are many situations where the microphone cannot be placed near the source of sound, for various reasons. These usually occur out-of-doors; bird songs, the crack of the ball on the cricket bat, the commands from the officer at a military parade are some examples. For these situations it is essential to have a microphone which picks up sound energy only over a small angle or to have, in other words, high directivity. This high directivity can be obtained by using interference effects; this is called the 'wave' type.

The simplest and probably the oldest example to provide this interference is the parabolic reflector. One can get an idea of how the reflector works by imagining the curved reflector facing a distant source on its axis. Energy coming from this source will approach the reflector as a plane wave. This wave will be reflected when it strikes the surface. It is a property of a parabolic surface that, irrespective of what part of the reflector it strikes, each part of the wave will arrive in the same phase at the focal point of the reflector. The path length from an axial source to the focal point via the surface is the same, irrespective of which part of the surface is used. Obviously at other angles of approach these conditions will not apply, and maximum sound intensity at the focus will only be obtained when the reflector is facing the source. Placing a pressure operated microphone at the focus, facing into the reflector, will produce a sound collecting device of high directivity.

Another method is to use a series of tubes, of small diameter and graded length, fixed in front of the diaphragm of a pressure operated microphone. When the tubes face the source, the contributions from each tube are arranged to arrive at the diaphragm in phase. For other angles of incidence, there will be a phase difference between the various contributions due to the different path lengths each has traversed. There will thus be maximum output when the tubes face the sound source.

A fundamental difficulty of the 'wave' principle is that, at low frequencies, there is little or no directivity with practical sizes of microphone. We have seen that no reflection takes place from an object unless the object is large compared with the wavelength. A practical size of reflector is 3 ft, which is effective down to about 800 c/s. Usually some form of bass cut filter is connected in the microphone's output circuit. Another point is that the reflector becomes very effective at high frequencies—short wavelengths—and there is extreme narrowing of the directivity pattern. This would cause serious loss of high frequencies if the source or the microphone moved. With some designs, this defect is corrected by introducing some defocussing—placing the microphone slightly away from the focal point. This broadens the high frequency directivity response. Other designs cover the outer part of the parabolic surface with absorbent material which is effective at high frequencies only. This reduces the effective size of the reflector and hence gives the same directivity at high frequencies as at low frequencies.

With the tube type of microphone, the directivity at low frequencies depends

TABLE 5.2. *Properties of Directional Microphones*

Type of Microphone	Approximate Size	Mode of Operation	Directivity Factor at Frequencies Indicated					
			200 c/s	500 c/s	1 kc	2 kc/s	4 kc/s	8 kc/s
Parabolic reflector type with omni mic.	3 ft dia. 1 ft deep Wt = 25 lb	Focal concentration of axial sound	4	20	47	125	—	—
Multi-tubular type. Staggered tube lengths	5 ft long 3 in dia. Wt = 5 lb	Phase interference of sound from staggered openings	3·5	8	12	25	25	40
Single tube with a damped slit	6 ft 6 in. long 1–2 in. dia. Wt = 4 lb	Phase interference of sound from different parts of slit entry	5·4	13	28	55	100	—
Single tube with discrete filter entries	6 ft 6 in. long Wt = 4 lb	Phase interference of sound from discrete entries along tube	6	15	30	50	45	60
Normal cosine or cardioid gradient mic.	Typically 1½ in. dia. × 4 in. long Wt = 1–2 lb	Pressure gradient operation	3	3	3	3	3	4

(Courtesy of S. T. & C. Ltd.).

on the maximum length; the larger this is the better the directivity, since there will be larger phase differences between the various contributions.

A modern version of the tubes idea is to allow the sound to enter a narrow slit running down a single tube which is coupled to a moving coil microphone. With this design, it is necessary to eliminate standing wave effects by having a layer of fabric over the slit. Another point is that sound energy entering at the front end of the slit will suffer loss in travelling down the tube. Energy entering further

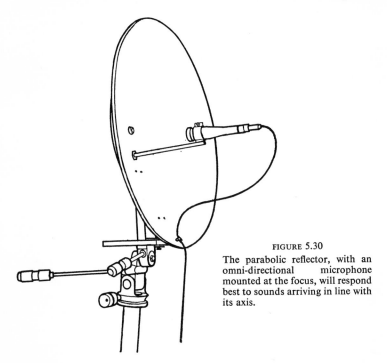

FIGURE 5.30

The parabolic reflector, with an omni-directional microphone mounted at the focus, will respond best to sounds arriving in line with its axis.

down the slit will not suffer the same loss, and this would lead to amplitude differences between the contributions instead of there being only phase differences. This is counteracted by having more absorbent material over the slit near the diaphragm. There is also some high frequency loss in coupling the slit to the microphone, and this is compensated for by having resonant cavities formed by small baffles fitted at 90° to the axis of the tube over the slit.

The tube or slit microphones are sometimes called 'line' microphones, and they have the advantage over the 3 ft parabolic reflector of weight and ease of handling. Smaller reflectors of 2 ft diameter are available for work where high directivity is not required at low frequencies. The parabolic reflector is more sensitive, however, since there is a gain due to the larger amount of sound energy being collected by the reflector. With the line type, the maximum sound which strikes the diaphragm when the microphone is facing the source cannot be more than would have struck it in any case. In fact, there is some loss of sound energy when the tubes or slot are connected in front of the microphone unit.

MICROPHONES IN PRACTICE

BEFORE beginning a short section on microphone placing, mention must be made of some effects which occur with microphones according to the way in which they are positioned.

In the section on ribbon microphones (p. 91), it was pointed out that, when using two ribbons close to each other, it is wise to check their phasing. Another point about using ribbons is that there is a very pronounced bass rise when the microphone is placed near to the source of sound. This occurs because the ribbon microphone is pressure gradient operated. In analysing the effect, two points must be noted:

1. When the source is close to the microphone, the wavefront of the sound wave striking the diaphragm is spherical. At normal distances, by contrast, the wave is virtually plane.
2. The driving force on the diaphragm increases with frequency, over the range being considered, when pressure gradient operation is being used.

Bass Rise with Pressure Gradient Microphones

In Chapter 1 we saw that the sound pressure in spherical waves falls off as the distance from the source increases. With plane waves the change in pressure for the same increase in distance is so negligible that it can be ignored. Now the difference in pressure between the two faces of the diaphragm comes about because the sound wave has to travel an extra distance—the path length—to get to the rear of the diaphragm. With plane waves this extra distance has no effect on the amplitude of the pressure acting on the rear face of the diaphragm; it has the same amplitude as the pressure acting on the front face. The only difference between the two pressures is one of phase. With spherical waves, however, the pressures now differ in amplitude as well as phase, since the pressure at the rear face will be reduced in amplitude owing to the wave having to travel an extra distance equal to the path length.

So we now see that reducing the distance from source to microphone below a certain value changes the type of the sound wave, and increases the difference in pressure between the two faces of the diaphragm. This increase in pressure difference occurs at all frequencies, but the reason why the effect is pronounced at low frequencies is the second of the two points mentioned previously—that the driving force increases with frequency.

TABLE 6.1. *Bass Rise in Pressure Gradient Operation*

Distance	Frequency	dB Rise relative to High Frequency Response
2 feet	50 c/s	6·0
	100 c/s	2·5
	200 c/s	1·0
	500 c/s	0
1½ feet	50 c/s	7·0
	100 c/s	3·0
	200 c/s	1·5
	500 c/s	0
1 foot	50 c/s	10·5
	100 c/s	5·0
	200 c/s	3·0
	500 c/s	0·5
6 inches	50 c/s	16·0
	100 c/s	10·0
	200 c/s	5·0
	500 c/s	1·5
	1000 c/s	1·0

At low frequencies, the difference in pressure between the two faces of the diaphragm is small and the driving force is thus small. At high frequencies, the difference in pressure is large and the driving force is thus large. Any increase in the pressure difference will therefore have a greater relative effect at low frequencies than at high. The driving force is therefore increased at low frequencies, while that at high

119

frequencies remains substantially the same and microphone output becomes bass heavy.

For normal use, the working distance for a ribbon microphone should be no less than 2 ft (but see Fig. 6.1).

Noise-cancelling Microphones

Although this rise in bass frequencies is a problem with the usual ribbon microphones, it is deliberately employed in one type to provide a most useful noise-cancelling microphone. This type is widely used in the broadcasting of commentaries from sports events where the background noise is high. The microphone is used very close to the mouth, the distance from the mouth to the ribbon being fixed by

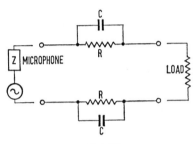

FIGURE 6.1

To correct the rise in low frequencies due to speaking too close to pressure gradient operated microphones such as ribbons, a circuit as shown in the sketch is used. It will also reduce the effect of room resonances which produce boominess. Typical values $C = 2\mu F$; $R = 5000\Omega$.

a bar on the microphone which rests against the upper lip. There is, as we would expect from the above discussion, a very pronounced rise in the bass frequencies. But this rise is corrected by equalisation, so that for the spherical waves from the close source the frequency response is flat. For plane waves from the surrounding 'noise' sources, there is heavy attentuation of the bass and middle frequencies. The microphone can thus be used in noisy surroundings to give very good signal to noise discrimination.

Effects on Close Speech

Using a close technique with sensitive microphones can cause difficulties owing to 'blasting' on explosive consonants such as 'p'. With pressure operated microphones, there is sometimes an increase in high frequency output when the source is on the microphone axis,

120

that is when the microphone is directly in line with the source. This is due to the microphone acting as a reflector at frequencies where its size becomes greater than the wavelength of the sound wave. The reflected energy interferes with the incident energy to form a standing wave system and, since the diaphragm is at a pressure anti-node, the output is increased. If this is found to give too much emphasis on the higher frequencies, altering the angle of incidence, by tilting or turning the microphone for example, will probably reduce it.

Microphones Out-of-Doors

Wind noises are a great source of trouble with outdoor broadcasting or recording. Normally ribbon microphones are useless out-of-doors, as the ribbon has to be fairly loosely tensioned and can be easily moved by wind. Moving coil and crystal microphones are suitable for outdoor use. If the wind noise is troublesome some form of wind shielding should be tried. Proper windshields are in fact supplied for some microphones, and these generally consist of layers of porous material which shield the microphone from the air currents but allow the sound wave to reach the diaphragm more or less unaltered. It is a good idea to insert some bass cut in the microphone circuit, as this will attenuate the turbulence tones.

Microphone Technique

Having discussed various fundamental topics about microphones, let us now consider how they can be used in obtaining the best results. Of course it is impossible to lay down exact rules which would produce perfect results every time. There are too many variables, the location, whether it is a speaker, a singer, etc., the microphones available and so on. However, it is possible to give some idea of the problems involved. Once these are understood, achieving a satisfactory balance becomes a matter of experience and patient experiment through knowledge.

Dynamic Range

Perhaps at this stage it would be a good idea to examine just what goes to make 'good quality'. First of all the microphone, amplifiers, tape recorders and loudspeakers should have a good frequency response. We have seen in a previous chapter that for music the range should be something of the order of 30–15,000 c/s. Two other important features the equipment should possess are a good signal/noise ratio and the ability to cope with a fairly wide intensity range. With music, as we saw earlier, the maximum range is approximately 70 dB, although this is not fully exploited by every kind of music. This is the intensity range of the source. To avoid, on the one hand, distortion due to overloading, and, on the other, noise due to low level, an intensity range of 40 dB or better is required from the equipment. If the demands for music are met, then all should be well for speech.

In broadcasting, the range has to be compressed further than this, to about 30 dB, and the 'light and shade' or 'contrast' of the programme is maintained by manual operation of the electrical controls. Carried out skilfully this control can be made inaudible to the listener. It is imperative that the person controlling the programme has a good knowledge of the material so that adjustments can be made at points where they are less noticeable.

So much for the equipment. If the requirements outlined above are met, the electrical signal fed into the chain will be handled without serious distortion. But now we must look at the acoustic signal which the microphone picks up, and the first point to remember is that most sources of sound do not distribute their energy in the same manner. Low frequencies are radiated uniformly, while high frequencies tend to be beamed in specific directions, so this point should be noted when positioning the microphone relative to the source.

Studio Acoustics

The next point is to see how the sound wave arriving at the diaphragm is made up, and this is where the acoustics of the room or hall come into the picture. Acoustics of rooms will be dealt with more fully in the next chapter of this book. For the purposes of this chapter, let us just consider that the total sound energy driving the diaphragm is made up of direct sound—from source to microphone; and reflected sound—which has struck one of the walls or the ceiling or floor of the room.

If the microphone is close to the source, there will be a high ratio of direct to reflected sound. If the microphone is placed well distant, the ratio will be low. Using a close technique will obviously stop the room acoustics having much effect, but the quality of the reproduction will be poor. As it lacks depth or perspective, the effect is unnatural. Coupled to this, of course, are the deleterious effects of working too near to the microphone; the bass rise with ribbons, for example, or the blasting on heavy passages. By a judicious placing of the microphone an acceptable ratio of direct to reflected can be obtained; near enough to retain clarity, far enough away to reproduce 'perspective'.

It has been remarked that the fundamental problem of acoustic design and microphone placing is to achieve satisfactory balance between direct and reflected sound. Broadly speaking, this is true for speech and for certain types of music, symphony orchestras and choirs for example. However, a close technique is often necessary nowadays and as an illustration one can consider the situation where a singer is singing quite softly against a loud dance orchestra. Here it is necessary to get some degree of separation between the singer and the orchestra, and the first step in getting this is to place the singer close to the microphone. To avoid the distortions, bass cut can be inserted if a ribbon microphone is being used and, if the singer turns her head, any excessive peaks can be avoided. In professional work, where condenser microphones in close balance are being used, they are fitted with special wind shields called 'close-talking shields' to avoid blasting on speech consonants.

Directivity of Microphones

The last item we are going to consider is the microphone and in particular its directivity pattern. In the preceding paragraphs, when discussing the influence of reflected sound in a room and the separation between a singer and orchestra, no mention was made of how the microphone responded to sounds coming from different angles. And yet it must be obvious that the directivity pattern of a microphone has a great bearing on the quality of the finished reproduction. An omni-directional microphone, for instance, will not discriminate against reflected sound and will produce a low ratio of direct to reflected sound. However, bi-directional or uni-directional microphones which have areas much less sensitive to sound can be angled to minimise reflected sound and will give better 'definition'. Directional microphones like ribbons and cardioids pick up about a third less reflected sound than omni-directional ones.

Use of the directivity patterns of microphones to cut out unwanted noises or room effects makes these microphones very versatile. In getting good separation, directional microphones are also invaluable, since the insensitive or dead face can be directed towards the unwanted source. And the directivity patterns are good tools for some dramatic effects. For instance, with a bi-directional microphone, moving from the live face into the dead side area gives the impression of the speaker having moved away into the distance.

A case where the development of robust cardioids such as the moving coil type has eased a practical problem is in public address work. Although directional loudspeaker systems are quite common,

123

the use of a cardioid microphone with its dead side towards the hall cuts down greatly the risk of the howl-round which occurs when the microphone picks up sound from the loudspeaker.

Illustrations of the basic principles discussed above are shown in the figures accompanying this chapter and they help to give some ideas in microphone balancing.

Stereophony

Up till now we have been dealing with the microphones and techniques used in monophonic systems. With good equipment and skilled use, the results obtained are very satisfactory. This has been the system used by broadcasting authorities and recording companies all over the world. And it has the merit of simplicity, the electrical output from a studio being available on one pair of wires, irrespective of the number of microphones used.

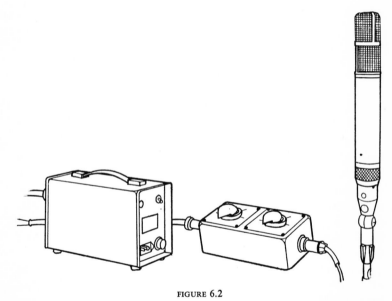

FIGURE 6.2

The A.K.G. C.24 stereo microphone is essentially two C.12 variable directivity condenser microphones in one unit. The supply unit and dual pattern selector are also shown.

But such a system cannot reproduce the whole of the sound image we perceive when we are listening to a source of sound directly. We have seen in Chapter 2 how our directional hearing abilities enable us to locate sources of sound and be conscious not only of the intensity of a source but also of its position. This we can do because of our two ears and our brain.

A monophonic system cannot recreate through a single loudspeaker the same acoustic impression which we would have received

FIGURE 6.3

The A.K.G. D.88 moving coil stereo microphone is popular with amateur tape recording enthusiasts.

had we been in the same position as the microphone. Many factors contribute to this discrepancy, but a considerable improvement in the naturalness of the reproduced sound can be obtained if we use a stereophonic system. The improvement leads to a sense of spaciousness when listening to a stereophonic system which the monophonic system lacks. This is due to more acoustic clues being fed ultimately to the listener, so that he is able to hear a better 'sound picture'.

Stereophonic systems start with more than one microphone; the exact number depends on which system is being used. The outputs of the separate microphones will differ depending on the distance and/or the angle of incidence of the sound waves to each microphone. When these outputs are amplified separately and reproduced on spaced loudspeakers, a directional effect is produced.

Although multi-channel stereophonic systems have been developed for the cinema, we will confine our attention to simpler methods. We will examine two microphone methods, first the 'wavefront' and secondly the 'coincident'.

125

Wavefront Method

The wavefront method was developed by engineers of the Bell System Laboratories in the United States during the thirties and was based on the theory that, if a series of microphones were arranged across the front of a stage and each one connected to a separate loudspeaker in an array arranged in the same way as the microphones, the wavefront of the sound coming from the loudspeakers would be the same as that striking the microphones. Practical difficulties would abound if a large number of microphones and loudspeakers were used, but by reducing them, an acceptable result was obtained.

Three microphones feeding their corresponding loudspeakers were tried, and this gave the listener sufficient acoustic clues for him to locate the position of the source across the stage. Of course, with a limited number of channels, when the source position was equidistant from two microphones, its reproduced effect was as if it had moved away from the microphones. Reducing the number of channels to two—i.e. a pair of microphones feeding two loudspeakers—aggravated this defect, which is often called 'hole-in-the-middle'. However, a two channel system is attractive from a simplicity point of view and two spaced microphones are frequently used for stereo today. The method can give very acceptable results, though in practice a third microphone is sometimes placed between the two others and its output connected equally between the two channels. This closes the gap somewhat.

Coincident Method

The other microphone technique used today is the 'coincident' one, where two directional microphones are mounted immediately adjacent to each other. Two spaced loudspeakers are again used for reproduction. This method was devised in Great Britain during the early thirties by A. D. Blumlein. The simplest way to understand it is to consider two figure-of-eight microphones so placed that their axes are at 45° on each side of the axial line to the source, i.e. the two axes are 90° apart (see Fig. 6.4).

When sound from the centre of the source strikes the microphones, equal outputs will be produced and the listener will locate the sound as coming from the middle of the acoustic sound stage between the two loudspeakers. If the source now moves to one side, a larger output will be produced by one microphone since the source has moved closer to its sensitive axis. This will produce a difference in the loudspeaker outputs, which the listener interprets as a move of the source to one side.

126

The coincident microphone method can give very good results, with positional accuracy of a high standard. Two figure-of-eight microphones have been mentioned, but other types can be used.

Another coincident microphone technique uses a cardioid and a figure-of-eight combined. The cardioid faces towards the source,

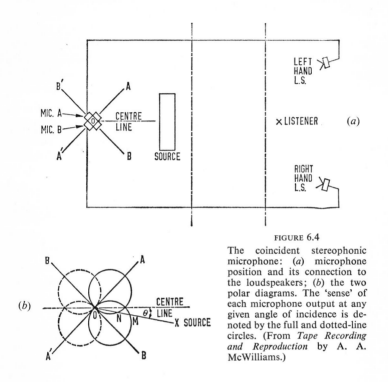

FIGURE 6.4

The coincident stereophonic microphone: (*a*) microphone position and its connection to the loudspeakers; (*b*) the two polar diagrams. The 'sense' of each microphone output at any given angle of incidence is denoted by the full and dotted-line circles. (From *Tape Recording and Reproduction* by A. A. McWilliams.)

while the figure-of-eight has one of its dead sides towards the source —it faces across the studio. In a variation of this method, the cardioid is sometimes replaced with an omni-directional type.

This method is called the M/S system. As far as the two output signals are concerned, the results it produces are the same as those obtained by the coincident method (see Fig. 6.5).

Microphones for Stereo

For professional work, condenser microphones are used consisting of two separate capsules with their associated head amplifiers inside the same case. This allows the two capsules to be placed close to each other. The upper capsule can be rotated from 0° to 180° relative to the lower system, and the directivity pattern of each capsule can be

127

individually altered. These facilities make very high quality results possible, since the variable controls allow fine adjustment of the microphone's response (see Fig. 6.2).

Naturally such complicated microphones are expensive, but there are now on the market a number of reasonably priced stereo types

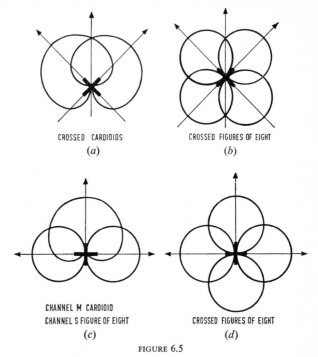

CROSSED CARDIOIDS
(a)

CROSSED FIGURES OF EIGHT
(b)

CHANNEL M CARDIOID
CHANNEL S FIGURE OF EIGHT
(c)

CROSSED FIGURES OF EIGHT
(d)

FIGURE 6.5

(a) and (b) show two methods of using coincident microphones in the X–Y system.
(c) and (d) show two methods of using coincident microphones in the M–S system.

for amateur work. These are usually of the moving coil or crystal type, having two cardioid units in one case. With such a microphone it is possible to get very acceptable results on many types of source. Orchestra, band, chorus, small instrumental groups, all produce quite good stereophony with one coincident microphone (see Fig. 6.3).

Limitations of Stereo

Unfortunately there are occasions when one wishes to reproduce more complicated sources and possibly use more than one microphone, and here it is important to be sure of the limitations of the stereophonic system.

128

As we saw earlier, a microphone's directivity pattern gives it a useful angle over which the sensitivity is fairly constant—this is sometimes called the 'acceptance angle'. With a stereo microphone, where the directivity patterns are angled at 90° to each other, the acceptance angle depends on what type of microphone is being used. With figure-of-eights it is approximately 80°; cardioids, as would be expected, give a wider angle of about 155° or 160°.

Let us examine what happens if we move a coincident microphone towards, say, an orchestra. If we begin at a considerable distance from the source, sound will be picked up by the microphones over only a small part of their acceptance angles.

The reproduced sound on our spaced loudspeakers will therefore appear to occupy only a part of the available sound stage width between the loudspeakers. If the stereo microphone is now brought nearer, the reproduced width will increase. Bringing the microphone too close will put the instruments at the side of the orchestra outside the useful angle of acceptance. This means that the central instruments will be reproduced satisfactorily, but the flanking instruments will be lost. So we can see that there is an optimum working microphone distance at which the reproduced sound will exactly occupy the full width of the sound stage.

Of course the effect of the room acoustics has been ignored in the foregoing simple argument, but it will be obvious that their effect will depend on the working distance of the microphone and this in turn will depend on the width of the sound stage.

The possible difficulties of having a solo singer with orchestral accompaniment, for example, can now perhaps be appreciated. If we place him near the stereo microphone to get adequate volume, the width of the reproduced voice will be unnaturally large compared with that of the orchestra. Taking him farther back, away from the microphone, will spoil the loudness balance between soloist and orchestra. A solution is to use a monophonic microphone on the soloist, and split its output between the two channels. This allows the soloist to be placed in some position away from the stereo microphone, so that he occupies a realistic part of the sound stage. And yet it can produce, through the monophonic microphone, an adequate output volume to give the correct balance between soloist and orchestra.

Obviously the apparent position of the soloist on the sound stage depends on the proportion of the signals fed to the two channels. If they are equal, he will appear to be in the middle. This underlines the need to match the position produced by the monophonic microphone and that produced by the stereo microphone. This is done by having faders which control the amount of the monophonic signal fed into each channel. These are called 'panoramic potentiometers' or simply 'pan-pots'.

If one is using two stereo microphones on two different sources of sound, it is again important to produce widths of sound stage which are acceptable. There would obviously be very considerable practical difficulties if one could only control the width by altering the microphones' working distances. The correct distance from a width point of view may not be that required for a good ratio of direct to reverberant sound.

Sum and Difference Technique

It would obviously be advantageous to be able to control the working width of any stereo microphone, and it is possible to do so by processing the two signals coming from each of the two units within one microphone. The signals are added to produce a *sum* signal, and increasing this decreases the width. They are also subtracted to produce a *difference* signal, and increasing this increases the width. By having ganged potentiometers controlling the sum and difference signals, one is able to adjust the width of the sound stage as required.

Microphone Balance

To illustrate how microphones can be used to the best advantage when recording different sources, some typical situations are shown and important points discussed.

Speech

Remembering that speech requires fairly 'dead' acoustics, the walls of the room in which speech is to be recorded should have some rudimentary absorptive treatment. A directional microphone, figure-of-eight or cardioid, should be used to increase the ratio of direct to reflected sound. To help in discriminating against some of the room resonances, the microphone should be placed along one of the room diagonals. Since these resonances produce standing wave patterns, relative movements of source and microphone can produce significant variations. Some trial positions should be recorded to get some idea of the room's acoustics.

The surface on which the microphone rests is important. If the surface is reflective, distortion can be produced by interference between this reflected sound and the direct sound so that partial cancellations take place in the sound reaching the microphone. In broadcasting studios, tables are made with special tops which are transparent to sound. A thick blanket thrown over the table in the living room will give some improvement compared with a hard, polished surface.

Now to the speaker; he should not sit too close—this point has been discussed earlier (p. 118). He must talk into the microphone and not look down at his script or move his head from side to side. He should not hold the script so that it obscures the microphone, otherwise 'shadowing' will occur at high frequencies.

When recording more than one speaker, the effect of the directivity pattern of the microphone should be remembered. In a play, for example, if one of the cast is standing near the 'dead' area of a figure-of-eight, it will sound as if he is standing at some distance from the rest.

Discussions and Interviews

The two sketches in Fig. 6.6 show how suitable arrangements can be made for recording interviews and discussions. If two ribbon microphones are used, the panel can be split so that, by suitable angling, two speakers can be put on the 'dead' face of the other microphone—and vice versa. This will avoid unwanted pick-up. A single cardioid can be used for this type of programme but care must be taken to ensure that all the speakers sound equally loud.

If the discussion is taking place in a public hall, a cardioid helps to avoid howl-round, but one cardioid may not be sufficient as the working distance has to be so reduced that the microphone only 'sees' the two middle speakers. Splitting the speakers into two pairs will obviously help here.

If an interview is being recorded with a pressure operated microphone, it is important to remember that at high frequencies the directivity pattern becomes very directional. If the interviewer talks directly into the microphone and forgets to turn it towards the speaker for the answer, there will be a considerable difference in quality between the two voices. It is better to hold the microphone at about chest level, equi-distant from each speaker and facing directly upwards, so that the angle of incidence is the same for both voices.

Musical Instruments

In Chapter 3 we saw that many musical instruments do not distribute the sound energy equally at all frequencies. High frequencies tend to become beamed, and contained in these frequencies are the upper harmonics which contribute a great deal to the individual instrument's tone quality.

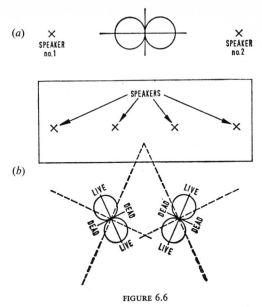

FIGURE 6.6

(a) Use of bi-directional microphone to balance two speakers.
(b) Two bi-directional microphones covering four speakers in line on stage. (From *Tape Recording and Reproduction*, by A. A. McWilliams.)

This is obviously a most important factor to be taken into account when recording a solo instrument. It is not the only one however; not only do instruments produce the required musical notes, but they also produce noises due to their mechanical action. Therefore care must be taken in placing the microphone to get a blend of satisfactory tone quality with the minimum of noise.

Strings

Of this family the violin has the most marked beaming of the higher frequencies, and the microphone should be placed at right angles to the front face of the instrument initially and then moved to one side perhaps until the noise is reduced. Obviously trial and error is called for, until the balance is right.

Of the other stringed instruments, it is important to let the cello 'see' the microphone. With its low playing position it is easy to lose the instrument among other instruments. With the double bass, one can often hear sounds which are just thumps instead of being clearly defined. This is more than likely due to the instrument causing resonances in the surroundings, and considerable improvement is possible by standing the instrument on a solid floor.

Brass Instruments

As with the violin, brass instruments can produce very directional high frequencies, and the same general considerations are important in recording them. The trumpet and the trombone are good examples of highly directional sound sources. Unlike the violin, however, they have a large output, so the microphone should be placed well away, say about five feet. If the player tends to move while playing, it is safer to place the microphone to one side of the axis of the instrument. The loss of some of the upper harmonics will not then be noticeable.

Piano

A great deal must obviously depend on whether the instrument is a grand or an upright, what type of music is being played and what the acoustic surroundings are like. With an upright piano, a reasonable balance is obtained with a bi-directional microphone placed as shown, it should be about three to four feet from the piano and two feet from the floor. Another way is to take the microphone round the front, and have it face the piano over the player's right shoulder.

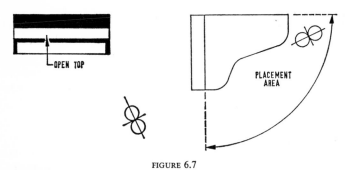

FIGURE 6.7

Illustrating the microphone placement for an upright piano (left) and a grand piano (right).

A grand piano, with the lid raised, radiates the sound energy fairly uniformly over a wide angle, and acceptable balances are possible over the area shown in Fig. 6.7. A directional microphone is usually

133

best facing the piano and arranged to be about six feet from the floor. Various working distances should be tried to suit the acoustics and the music being played. If it is dance music, a close position should be used; but with much classical keyboard music there is a wide range of intensities and the microphone should be moved further back.

Using the piano as a member of the rhythm section of a dance band presents certain problems, one being to get a good 'tight' sound free from reverberation and the sounds from other powerful instruments. The technique used in broadcasting and recordings is to place a microphone close to the treble strings, and remove the lid so that no reflections can take place.

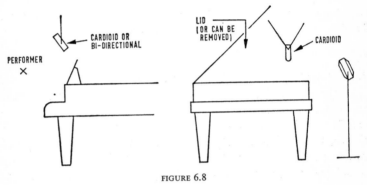

FIGURE 6.8

A single cardioid microphone may be used for songs at the piano (left). Alternatively, a second microphone may be introduced for the piano (right).

Woodwind Instruments

A satisfactory balance is obtained fairly easily with most woodwind instruments since they are not particularly directional. A microphone about two feet away and placed so that the instrument does not 'speak' straight at it should give acceptable results.

Percussion

Even in professional work the microphone placement for these instruments is difficult since they tend to be powerful and call for a distant microphone to cope with the volume. However, this means that the result is 'blurred' or lacking in definition. So with drums and cymbals various microphone positions should be used until a satisfactory compromise is obtained.

134

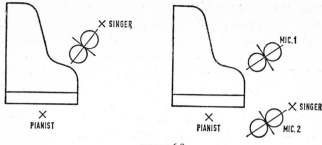

FIGURE 6.9

For singer with piano accompaniment, a single bi-directional microphone is suitable (left). If a tighter balance is required, two microphones may be used, suitably angled to pick up singer and piano
individually (right).

Songs with Piano

With an upright or a grand piano, a single microphone technique
gives good results with the singer on one side and the piano on the
other. Working distances should be adjusted to take care of the
relative strengths of the singer and piano. One situation to avoid is
allowing the singer to stand in the 'bend' of the grand piano; this
makes for poor results.

Singers and Choirs

Mention has been made of the dangers of a solo singer coming too
close to the microphone, and precautions should be taken against
the possibility of blasting either by getting the singer to turn his head
slightly or by windshielding the microphone.

Small groups of say four singers can be arranged on either side of
a bi-directional microphone to give a very close balance.

Or, of course, they can be placed together and a cardioid used. If it

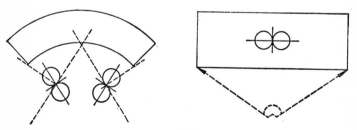

FIGURE 6.10

Very large choirs or other groups may require two bi-directional
microphones to cover them effectively (left). If a cardioid is used, a
bi-directional microphone may be introduced as shown to add
'presence' (right).

135

is a large group, good results can be obtained by slinging the microphone above the group and arranging that it 'looks' downwards.

With vocal groups a fairly close balance with little reverberation is required as this gives good diction. However, with choirs much of the music they perform was written to be sung in reverberant surroundings and consequently a distant microphone position is required consistent with retaining acceptable diction.

A cardioid is useful since it allows the choir to remain as one unit. A ribbon would achieve the same purpose but if the choir is large some of the singers will be outside the acceptance angle. Of course, if it is possible, the choir can be split into two parts and placed on either side of the ribbon microphone.

Organ

Here it must be remembered that church organs were designed to be played in reverberant conditions and much of the beauty of organ music lies in the full sound which it produces. Any balance therefore must be able to reproduce this, so a distant microphone position is required. Care must be taken however to ensure that the clarity of some of the notes is not lost and it may well be found that the layout of the pipes demands more than one microphone.

ROOM ACOUSTICS

WHEN a source is radiating sounds in a room, the resultant sound 'picture' we receive depends very much on the characteristics of the room. The size of the room, how it is constructed, the furnishings, etc., all have a bearing on how the room reacts to the sound waves coming from the source. Of course, when we listen to the sound directly with our two ears, we are able to discriminate against a great deal of the room's effects. Since we have binaural hearing, this enables us to reject sounds, to a great extent, which are not coming straight from the source. A microphone cannot do this since it merely measures the total sound pressure at one point in space, and cannot discriminate between the direct sound from the source and sounds arriving by other paths.

This, of course, refers to single channel or monophonic technique; we saw earlier how a stereophonic system is able to pass more information to the listener (see p. 124). This increase in information about the sound pressures in a room allows the listener to build up a much more realistic acoustic picture.

But for the moment let us consider the monophonic case; because a very satisfactory standard of reproduction can be obtained if one understands the problems involved. A brief mention of the importance of room acoustics was included in the last chapter (see p. 122) where we saw how controlling the ratio of direct to reflected sound, by microphone techniques, affected the clarity of the reproduction. Now let us see how these direct and reflected sounds behave in a room.

How Sound Behaves in a Room

Consider a source of sound, which we can switch on and off at will, placed a little way from the microphone, which in turn is situated near the centre of the room. When the source is switched on, the first sound energy to reach the microphone will be that which travelled

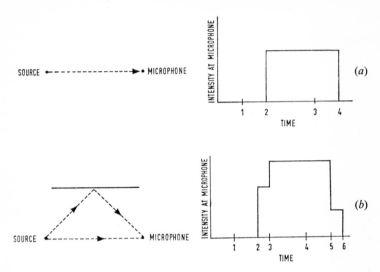

FIGURE 7.1

In (a) sound is assumed to travel by the direct path only. When the source is switched on at time 1, the sound wave from the source takes until time 2 to reach the microphone. When it does so the intensity at the microphone rises to a maximum value. At time 3 the source is switched off and this causes the intensity at the microphone to fall to zero at time 4.

In (b) a single reflected path is added. Energy travelling by this path arrives at the microphone at time 3 and will raise the total intensity to a higher value than in (a). The effect of the reflected path is also to delay the fall of the intensity until time 6.

by the direct path. The next to arrive will be that due to first reflections; the sound will have been reflected once by the boundaries of the room. More than likely there will have been some loss of sound energy on reflection, so that the energy arriving by this path will be less than that arriving by the direct path. The total energy received will therefore be greater than if no walls were present (see Fig. 7.1).

Next to arrive will be the energy which has been reflected twice; the second reflections, having struck two of the boundaries, will have suffered more loss than the first. The energy will increase again.

It is now apparent that as time increases, from the moment of switching on, the energy at the microphone increases as more energy

138

comes in by the reflected paths. Of course the contributions from the reflections decrease as time goes on, since the later reflections will have struck the reflecting surfaces more often and will accordingly have suffered more loss. After a time, contributions from later reflections will be so small that they can be ignored and the received energy will have reached its maximum value.

Equilibrium Intensity

As there are many reflections in a room, the energy at the microphone does not increase in discrete steps. It tends to be a gradual rise, steeper at first and then gradually levelling out till it reaches its maximum value (see Fig. 7.2).

The intensity at this maximum value is called the equilibrium intensity. The source is supplying energy at a certain rate into the room and the sound waves lose energy on being reflected. When the rate at which energy is being lost equals the rate at which it is being supplied, the energy, or intensity, of the sound in the room is constant and this value is the equilibrium intensity.

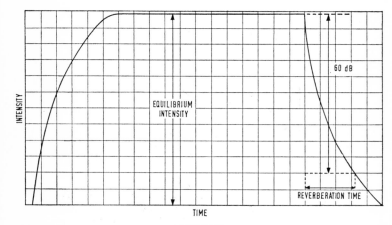

FIGURE 7.2

The variation in intensity with time in a room. This is an idealised 'smoothed' curve. In practice the intensity fluctuates about this curve due to interference effects between the various sound waves making up the total intensity.

The effect of the reflections from the wall is therefore to increase the intensity of sound available from the source. In practice, this is obviously important in getting adequate volume in a given room, without strain on the part of the speaker or musician.

Reverberation

Now let us switch the source off. The effect of doing so will be felt first by the direct path, since it is the shortest, and the energy contributed by the direct sound will be the first to disappear. Next the contributions due to the first reflections will disappear, then these due to the second reflections and so on. The intensity of sound gradually decays from the time that the source was switched off. It is clear that the sound in the room does not immediately fall to zero when the source stops radiating, but is prolonged for some considerable time. This prolongation is called reverberation and it is a most important factor in room acoustics. The presence of reverberation, and its time and frequency characteristics, largely control the sound quality of a room (see Fig. 7.3).

Any effect the reverberant energy has will depend on the length of time for which it is present. In a room with highly absorbing surfaces, the energy contributed by reflections will be low and there will not be a great deal of reverberation. If the surfaces could be made less absorbent, the reverberant energy would be greater and the sound

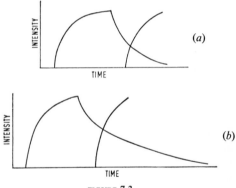

FIGURE 7.3

(a) The behaviour of sounds in a room having a short reverberation time. It can be seen that by the time the second sound rises the intensity due to the first sound has fallen to a low value.

(b) In a room having a long reverberation time, the second sound builds up when there is still considerable intensity due to the first sound. If these were speech syllables it can be appreciated how this overlapping causes poor intelligibility.

in the room would continue to be audible for a longer time. The difference in sound quality would be immediately obvious. Therefore, by controlling the effectiveness of the absorption present in the room, we can control the rate at which the reverberant energy dies away. That is to say, we can control the Reverberation Time.

140

The reverberation time is the most important single parameter of a room. It is defined as the time taken for the sound intensity to fall by 60 dB from the equilibrium intensity. This definition is used when

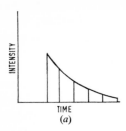
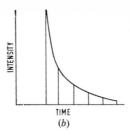

FIGURE 7.4

Since reverberation is due to reflected sound, its effect will depend on the ratio of direct to reflected sound. At a normal working distance the reflected sound will form a considerable part of the sound picked up by the microphone. If, however, the working distance is reduced, the direct sound will rise in intensity and hence the ratio of direct to reflected sound will be increased. Sketches (a) and (b) represent the two conditions. From this it is obvious that quite a different reverberant quality can be produced by varying the microphone distance.

stating the measured reverberation time of a room and for the moment we shall accept it; but later we will examine *how* the energy in the room falls, not just the extent of the fall.

Effect of Room Volume

Now let us look at the other factors which control the reverberation time besides the absorption of the surfaces. In fact there is only one important feature, the volume of the room. We have seen that, as the absorption is increased, the reverberation time falls. To this must be added the fact that the reverberation time increases if the room volume is increased. These three factors are related in a formula established by the American physicist, W. C. Sabine:

$$T = \frac{0{\cdot}05 \ V}{A}$$

where T = reverberation time in seconds,
V = volume of room in cubic feet,
A = total absorption in room

Optimum Reverberation

This relationship shows that, with a given volume, the reverberation time is determined by the total absorption present. The question now arises as to what value of reverberation time is required, and here we enter into the realm of opinion and almost fashion. The effect of

141

reverberation, providing extremes are avoided, depends on the subjective judgment of an observer and it is consequently impossible to lay down hard and fast rules which will yield universally acceptable reverberation times. However, widely acceptable empirical data has been accumulated over the years and from this an acoustic consultant can obtain an optimum reverberation time for the volume of the room or hall he is designing. Then the required absorption can be calculated. This gives a guide to the types of wall and ceiling treatment to be applied. Some adjustment is usually necessary during the final stages of construction to obtain the best results.

Some typical reverberation times are as follows (see also Table 7.1):

Domestic Living Room	0·35 sec.
Broadcast Talks Studio	0·3 sec.
Theatre	1·0 sec.
Medium-sized Broadcast Music Studio	1·2 sec.
Large Broadcast Music Studio	1·8 sec.
Large Concert Hall	1·8 sec.

These figures reflect the fact that a fairly short reverberation time is necessary for speech, otherwise listening conditions become difficult. This is because speech is made up of syllables and, for maximum intelligence, each syllable should be easily discernible. If the rever-

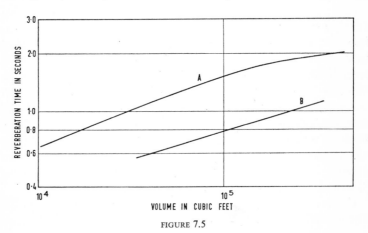

FIGURE 7.5

The optimum reverberation time curves for BBC studios. Curve *A* refers to music studios in the Sound Service; Curve *B* is for general purpose television studios. (By courtesy of BBC.)

beration time is long, each syllable will be so prolonged that its decay interferes with the succeeding syllable and the sound becomes confused and difficult to follow.

Reverberation and Music

For music, however, fullness of tone is required both by the audience and the performers. It is essential that the players hear themselves

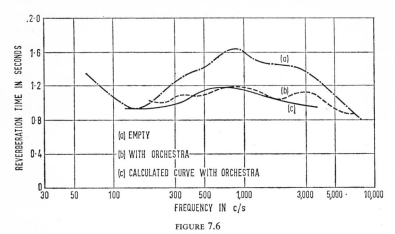

FIGURE 7.6

The reverberation time/frequency curves for a small orchestral studio. (By courtesy of BBC.)

and each other easily. If they cannot, they tend to force the tone to obtain adequate volume and ensemble playing suffers. These effects are apparent to the audience and, with the absence of reflections farther down the hall, the final result is said to be 'dry'; the music lacks resonance. Therefore, to obtain the degree of reflection which is required, auditoria for music should have a fairly long reverberation time.

Another desirable quality for music is good definition or clarity and, recalling the previous paragraph on speech, this would be obtained with a short reverberation time. An acceptable compromise has therefore to be secured by the designer, to reconcile these two differing requirements—a fairly long reverberation time for fullness of tone, and a short one for good definition.

Of course, a great deal depends on the type of music being played. The acoustic requirements for Mozart, for example, are different to those for Rachmaninov. For Choral music a longer time is preferred than that for orchestras. In Cathedrals and Chapels the reverberation time is usually several seconds.

Up till now, we have only been considering time in relation to reverberation and taking no account of frequency. Obviously, however, information on the behaviour at different frequencies is required.

143

This is best shown in a graph plotting the reverberation time at all frequencies in the audio band. The question arises as to the ideal shape for this graph. There is no universally accepted shape, various authorities differing in their ideas, but in general a reasonably flat graph is the aim. Excessive rise in the reverberation time at one frequency, or over a band of frequencies, is usually regarded as a bad defect in the acoustic performance of a room or hall. For example, if the low frequency reverberation time in a concert hall is high with respect to the other frequencies, there is a tendency to boominess due to the bass instruments receiving considerable re-inforcement from the acoustics of the hall. In small rooms, such as those used for speech studios, there can be difficulty with the reverberation time rising considerably at particular frequencies. This is an example of a defect usually called coloration.

Effect of Audience and Performers
In designing for certain reverberation time in a concert hall, the absorption effect of the audience must be borne in mind and due

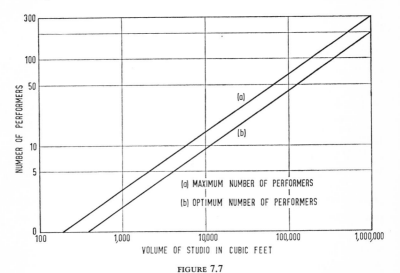

FIGURE 7.7
The performance of a studio depends on the number of performers in the studio. The graph relates the number of performers and the studio volume. (After Kirke & Howe; by permission of I.E.E.)

allowance made. With orchestral broadcasting studios the effect of the orchestral players must be noted and indeed there is definite relationship between the size of a studio and the maximum number of players it should accommodate. Published graphs show how these effects alter the reverberation time, and how, for example, the reverberation time when the hall is empty differs from that obtained when the audience is present. To minimise the effect of variations in the size of audience, specially designed seating is often installed, so that unoccupied seats provide some absorption.

More Exact Formula

The foregoing pages have given a very much simplified account of how sound behaves in a room, and about reverberation time and its

TABLE 7.1. *Reverberation Times*

Studio or Concert Hall	Volume in cubic feet	R.T. in seconds
Talks Studio	2,000	0·3
Light Entertainment Studio	60,000	1·0
Light Music Studio	113,000	1·5
Music Studio	180,000	1·7
Large Music Studio	220,000	1·8
Royal Festival Hall, London	775,000	1·47
Free Trade Hall, Manchester	545,000	1·6
Usher Hall, Edinburgh	565,000	1·65
Concertgebow, Amsterdam	663,000	2·0
Beethovenhalle, Bonn	555,340	1·7
Grosser Musikvereinssaal, Vienna	530,000	2·05
Carnegie Hall, New York	857,000	1·7
Symphony Hall, Boston	662,000	1·8
Royal Opera House, London	432,500	1·1
La Scala, Milan	397,300	1·2
Metropolitan Opera House, New York	690,000	1·2

The reverberation times are mean values over a frequency range from 500–1,000 c/s.

The concert hall and opera house figures are with full audience and are quoted from *Music, Acoustics and Architecture*, by Leo Beranek. Published by John Wiley and Sons Inc.

control. The account is based on several assumptions which are not true for all conditions. One might ask—why make these assumptions? The answer is simply that if one tries to be more accurate, the analysis gets so complicated that it is of little practical value. Thus the usual procedure is to make these assumptions to arrive at a suitable design, and then make corrections based on practical experience during construction.

One assumption made was that all the reflections arriving at the microphone are in phase. This meant that reflections would arrive and add to the energy already present, and the rise to the equilibrium intensity could be assumed to be smooth. Similarly, the decay could be taken as smooth, because the reflections would die out both in amplitude order as well as time order. In practice, however, this smooth pattern very rarely occurs. Both the build-up and decay are irregular due to phase and amplitude variations. But the patterns approximate to the curves which the simplified approach predicts.

Again, Sabine's formula resulted from assuming that the sound energy distribution is uniform throughout the room and that it would not matter where the measuring point was. This is termed a perfectly diffuse sound field. It is probably a most desirable acoustic property but one which is difficult to achieve in practice. Other assumptions cause slight errors in the reverberation time calculations using Sabine's formula. The overall error is found to be negligible with large volumes such as concert halls, but it increases when the amount of absorption is high, as is usual in small rooms. A modification to the Sabine formula gives a more accurate result in the latter case. This modification is due to Eyring, and gives the expression worked out in the following section.

Eyring's Modification
The complete Sabine formula for reverberation time is

$$R.T. = \frac{0.05\,V}{S\bar{\alpha}} \text{ seconds,}$$

Where V is the volume of the room, S the total surface area and $\bar{\alpha}$ the average absorption co-efficient.

Absorbent co-efficients will be discussed later in the chapter. At this stage we will merely note the fact that they range from 0 to 1·0, being 0 when the absorption is zero and 1·0 when there is complete absorption.

In a room there will be several areas of different materials and construction having different co-efficients. The total absorption is given by:

$$A = \alpha_1 S_1 + \alpha_2 S_2 + \alpha_3 S_3, \text{ etc.,}$$

the average absorbent co-efficient is given by:

$$\bar{\alpha} = \frac{\text{Total absorption in absorption units}}{\text{Total surface area}} = \frac{A}{S}$$

Sabine's formula suggests that, if $\bar{\alpha}$ was 1·0, the reverberation time would have a value which varied according to the dimensions of the room. But 100% absorption would mean that there were no reflections and hence no reverberation—one would have a completely 'dead' or an-echoic room.

This error is due to assumptions made in the theory leading to the formula. One assumption has been mentioned above—that the sound field is perfectly diffuse. Another is that the absorption of sound is a continuous process. But Eyring reasoned that the absorption would take place in discrete steps, when the sound waves struck the absorbent surfaces. The absorption process was thus a discontinuous one which would depend on the time interval between successive impacts.

This time interval would in turn depend on the path lengths the sound waves had to travel between impacts. To meet this, the concept of a 'mean free path' was introduced and this is given by $4\dfrac{V}{S}$, where V and S are as above. At each surface the reflected energy is $1 - \bar{\alpha}$, and the average number of reflections in time t is $\dfrac{Sct}{4V}$, where c is the velocity of sound in air.

If during the period t the intensity drops by 60 dB, the reverberation time formula works out as:

$$R.T. = \frac{0.05\ V}{S \log_e \left(\dfrac{1}{1 - \bar{\alpha}}\right)} \text{ secs.}$$

$$= \frac{0.05\ V}{-S \log_e (1 - \bar{\alpha})} \text{ secs.}$$

When $\bar{\alpha}$ is small, say below 0·3, $\log_e (1 - \bar{\alpha})$ is practically equal to $\bar{\alpha}$ and so, when the co-efficients are small, the simple relationship shown below can be used:

$$R.T. = \frac{0.05\ V}{\alpha_1 S_1 + \alpha_2 S_2 + \alpha_3 S_3, \text{ etc.}} \text{ secs.}$$

When $\bar{\alpha}$ is above 0·3 a more accurate result is obtained by using Eyring's modification as quoted above, and it is widely applied in the design of studios for broadcasting and films.

Room Resonances

Before discussing this modification in detail, it is pertinent at this stage to examine one respect in which a small room presents more difficult acoustic problems than say a concert hall. We saw in Chapter 3 how an air column can be made to resonate with a source of sound if its length bears a certain relationship to the wavelength of the sound. When resonating, the air column has a standing wave pattern, and it is possible to set down the pattern distribution provided that

147

the conditions at the end of the air column are known. With a pipe open at both ends, the fundamental note is produced when the length is half the wavelength and there is an anti-node of displacement at each end (see Fig. 7.8*a*).

Now imagine that we close both ends, but that we can still manage to excite the air by some enclosed source. There will obviously be

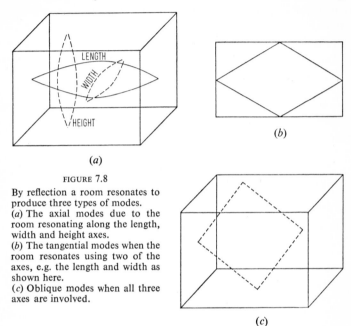

(*a*)

(*b*)

(*c*)

FIGURE 7.8

By reflection a room resonates to produce three types of modes.
(*a*) The axial modes due to the room resonating along the length, width and height axes.
(*b*) The tangential modes when the room resonates using two of the axes, e.g. the length and width as shown here.
(*c*) Oblique modes when all three axes are involved.

reflections from the closed ends, and a standing wave system will be set up. The length/wavelength relationship will be the same as for an open pipe, but the pattern distribution will be altered in that there will now be displacement nodes at each end. But a rectangular room can be regarded acoustically as being the same as a closed pipe. The room will resonate if its length is half the wavelength for some frequency present in the sound source. Unlike the pipe, a room can obviously resonate in its other dimensions, behaving like a three-dimensional resonating box.

The air in a room can vibrate and resonate in a great many different directions. Normally these are called the 'modes' of a room. First of all there are the axial modes, the resonances due to each of the axes of the room—length, width and height. Then there are modes which use two of the axes, for example the length and the width; these are called the tangential modes. Finally, there are those which use all three dimensions, the so-called oblique modes. All these types will, like the air column, support a fundamental and its harmonics, the lowest fundamental being due to the longest axial dimension (Fig. 7.8).

Axial Modes

It is interesting to calculate these room resonances for a typical small room, having the following dimensions:

$$\text{Length} = 15 \text{ ft}$$
$$\text{Width} = 13 \text{ ft } 6 \text{ in.}$$
$$\text{Height} = 8 \text{ ft}$$

For simplicity, we will start by calculating the axial modes. To do this, we will use the formula derived in Chapter 3 which shows that the fundamental note of an air column and its length are related thus:

$$\text{Fundamental} = \frac{c}{2l} \text{ c.p.s.}$$

where c = velocity of sound in air = 1,120 ft/sec.

Length Mode

$$\text{Fundamental} = \frac{1,120}{2 \times 15} = 37 \text{ c.p.s. approx.}$$

Width Mode

$$\text{Fundamental} = \frac{1,120}{2 \times 13 \cdot 5} = 42 \text{ c.p.s. approx.}$$

Height Mode

$$\text{Fundamental} = \frac{1,120}{2 \times 8} = 70 \text{ c.p.s. approx.}$$

From these three results, we can see why room resonances, or *eigentones* as they are sometimes called, are so important since these frequencies of resonance lie well inside the audio range.

Other Modes

Having found the axial modes, let us now see how the tangential and oblique modes are calculated. The simple formula we used above was only applicable to a one dimensional system. We must modify it for the two-dimensional tangential modes and the three-dimensional oblique modes. In fact, the usual method is to use the general formula by which all the modes can be found:

$$f = \frac{c}{2}\sqrt{\left(\frac{p}{l}\right)^2 \quad \left(\frac{q}{w}\right)^2 + \left(\frac{r}{h}\right)^2} \text{ c.p.s.}$$

The length, width and height are denoted by l, w and h; c is the velocity of sound as before; p, q, r, can be any whole number 0, 1, 2, 3, etc., and these are known as the mode numbers. For example, the fundamental in the length direction was the $1:0:0$ mode and the frequency was strictly calculated as follows:

$$f_L = \frac{C}{2}\sqrt{\left(\frac{1}{l}\right)^2 + \left(\frac{0}{w}\right)^2 + \left(\frac{0}{h}\right)^2}$$

$$= \frac{C \times 1}{2 \times l} \text{ c.p.s.}$$

The second harmonic for the length is the $2:0:0$ mode; the fundamentals for the width and height are the $0:1:0$ and $0:0:1$ modes respectively. The fundamental for the tangential mode using the length and width is denoted by $1:1:0$, and the fundamental oblique mode is $1:1:1$. One can see that if there are two zeroes for p, q or r, an axial mode is being considered; if there is one zero, it is a tangential mode, and if there are no zeroes, it is an oblique mode.

Of course, when we record a person speaking in a room, for example, three factors determine the sound pressures which operate the microphone:

1. The frequencies and amplitudes of the speech.
2. The acoustic properties of the room.
3. The directional properties of the microphone.

Effect on Sound Quality

The characteristics and frequency ranges of speech were outlined in Chapter 4. In simple terms, we have now seen how the air in a room vibrates and resonates, and how the resonant frequencies can be calculated. Now let us consider the relative importance of the different modes, and the effect of the resonances on the acoustic performance of the room.

The axial modes are the most important. They contain most of the reverberent energy and hence they determine the reverberation time. The effect of a particular mode depends on its bandwidth and how far it is away from the adjacent modes. The bandwidth obviously determines the chances of the mode being set into resonance. If it does resonate, and it is fairly isolated, then there will be considerable reinforcement of the original sound by the room at the resonant frequency; and the resultant strong coloration will cause annoying boominess. It is therefore highly desirable to make the axial modes equally spaced throughout the important low frequency range, so that no mode is isolated. Since the mode frequencies depend on the room dimensions, control of their spacing can be achieved by a suitable choice of length, width and height. This is done when talks studios or auditoria are being designed.

Improving Room Acoustics

In domestic rooms one has, of course, to make do with the room size and shape as it exists. However, one can improve the quality by using a directional microphone such as a ribbon. This can be placed or angled so as to suppress the coloration due to certain dimensions. Placing the microphone along one of the diagonals of the room will reduce the effect of the axial modes. A cardioid will give a similar improvement. Another means of improvement is to insert some bass cut in the microphone circuit. This is often done in professional studios, and it is worth trying if the speech reproduction is marred by 'boominess'.

Another factor in controlling room resonances is the distribution of the absorption materials. The materials are most effective if they are broken up into relatively small patches and randomly distributed. Although this is comparatively simple in a talks studio, it is more difficult in a room in the home; but if it is possible to have some absorption material on each wall, this will help towards better results.

Sound Absorption Methods

All absorbents work by dissipating acoustical energy as heat or mechanical energy. There are now many materials and methods used in practice by architects, broadcasting organisations, etc., but one can divide them into two main categories, porous absorbers and resonant absorbers.

Porous Absorbers

With porous materials such as felt, glass wool and acoustic plaster, the material contains air in a network of interconnected cavities. When the sound waves set up air vibrations, heat is generated due to friction within the material and the sound wave loses energy. The absorption of a material depends on the area of the material which is contacted by the sound waves, and its absorption co-efficient. The effect of a material on a sound wave will obviously depend on how

151

much sound energy strikes it, in other words on the surface area of the material. The absorption co-efficient is a measure of how efficient the material is as an absorber; a co-efficient of 1·0 would represent a material which absorbs all the sound energy striking it; a co-efficient of 0 represents a material which reflects all the energy. Thus practical absorption co-efficients lie between 0 and 1·0. If this is taken as the absorption due to each square foot of surface area of the material, the product of the surface area and the absorption co-efficient gives the absorption in convenient sq. ft units (see Table 7.2).

Porous materials have co-efficients which can vary widely with frequency. Taking two examples, the absorption co-efficient for a carpet with a felt underlay laid on the floor varies from 0·15 at 250 c.p.s. to 0·75 at 2,000 c.p.s., and that for glass wool varies from 0·25

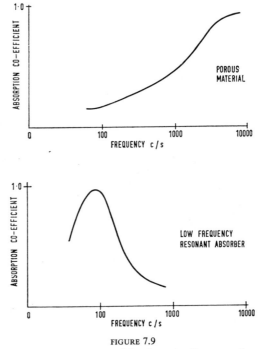

FIGURE 7.9

Comparing the general shape of the absorption/frequency characteristics of porous materials (top) and resonant absorbers (bottom).

to 0·7 over the same frequency range. It will be noted that the absorption is generally much greater at the higher frequencies (see Fig. 7.9).

The co-efficient also depends on the thickness of the material used. With the glass wool, for instance, the figures quoted above were for

TABLE 7.2. *Absorption Co-efficients*

Material	Thickness	Frequency in cycles/second						
		125	250	500	1,000	2,000	4,000	8,000
Brickwork (plain or painted)		0·05	0·04	0·02	0·04	0·05	0·05	
Coke Breeze Blocks	3 in.	0·2	0·45	0·6	0·4	0·45	0·4	
Acoustic Plaster	1 in.	0·05	0·1	0·15	0·15	0·2	0·25	
Sprayed Asbestos	1 in.	0·2	0·35	0·6	0·75	0·75	0·75	
Glass-wool	2 in.	0·2	0·45	0·65	0·75	0·8	0·8	
Glass-wool	4 in.	0·45	0·75	0·8	0·85	0·9	0·85	
Glass-wool—Resin Bonded	1 in.	0·1	0·35	0·55	0·65	0·75	0·85	0·75
Glass-wool—Bitumen Bonded	1 in.	0·1	0·35	0·5	0·55	0·7	0·7	0·75
Mineral Wool Tiles 12 in. sq.	¾ in.	0·1	0·2	0·5	0·85	0·85	0·85	
Flexible Polyurethene Foam	2 in.	0·25	0·5	0·85	0·95	0·9	0·9	
Rigid Polyurethene Foam	2 in.	0·2	0·4	0·65	0·55	0·7	0·7	
Expanded Polystyrene (on 2-in. thick battens)	1 in.	0·1	0·25	0·55	0·2	0·1	0·15	
Perforated Fibre-board (Tiles on 2-in. thick battens)	2·4 cm	0·3	0·45	0·5	0·55	0·65	0·8	
Rockwool (covered by Hardboard, ⅛-in. thick open 5% area)	2 in.	0·35	0·75	0·85	0·7	0·45	0·25	
Axminster Carpet	0·3–0·35 in.	—	0·05	0·15	0·3	0·45	0·55	
Turkey Carpet	0·6 in.	—	0·1	0·25	0·5	0·65	0·7	
Axminster Carpet (on needleloom underfelt)	0·5–0·6 in.	—	0·15	0·4	0·6	0·75	0·75	
Cork Tiles	$\frac{9}{16}$ in.	—	0·05	0·15	0·25	0·25	—	
Sorbo-Rubber (with thin layer of hardwearing surface)	$\frac{5}{16}$ in.	—	0·05	0·15	0·1	0·05	—	
Plate Glass Windows	¼ in.	0·1	—	0·04		0·02	—	
Audience-absorption units per person when seated in fully upholstered seats		2·0	4·3	5·0	5·0	5·5	5·0	

The absorption co-efficients were obtained by reverberation method and are quoted from *Sound Absorbing Materials*, by Evans and Bazley. Published by H.M. Stationery Office. This is an excellent, inexpensive booklet on the subject.

a 1-in.-thick layer with a rigid backing. If the thickness is now doubled, the co-efficient becomes 0·45 at 250 c.p.s. and 0·8 at 2,000 c.p.s.

We can see why the thickness affects the absorption efficiency by recalling the behaviour of standing waves as explained in Chapter 1 (see p. 24). When a wave is reflected from a rigid surface, the air particles have zero velocity (node) at the surface, and the first

153

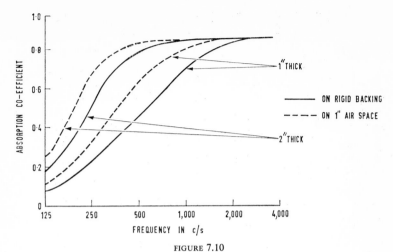

FIGURE 7.10
Showing how the absorption co-efficient of a porous material varies
with thickness and method of mounting. (From *Sound Absorbing
Materials*, Evans & Bazley. H.M.S.O.)

position of maximum (anti-node) velocity is a quarter of a wavelength
away from the surface. Now the absorption of sound energy in a
porous material depends mainly on the particle velocity of air enclosed inside the pores. At low frequencies—long wavelengths—the
distance of the anti-node from the reflecting wall will be large, and the

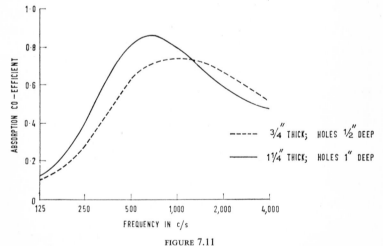

FIGURE 7.11
The effect on the absorption co-efficient of the depth of holes and
thickness of perforated fibreboard tiles. The holes act as resonators
and their depth controls the resonant frequency. (From *Sound
Absorbing Materials*, Evans & Bazley. H.M.S.O.)

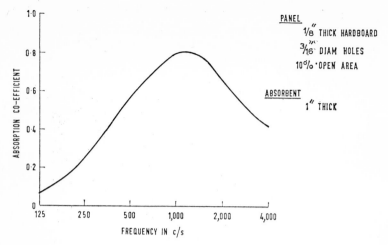

FIGURE 7.12a

Perforated hardboard is widely used as a covering for porous materials to reduce the absorption of the layer at high frequencies. The above graph shows how the absorption co-efficient falls above 1000 c/s for the particular materials used.

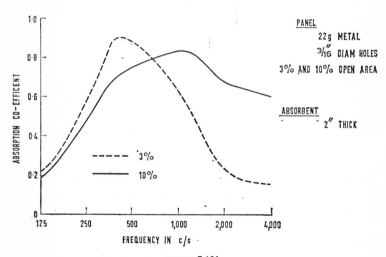

FIGURE 7.12b

The effect of a perforated covering depends on its thickness and the percentage of open area which in turn depends on the size and spacing of the holes or slits. This can be seen in the above graph which shows the absorption co-efficient for a perforated metal covering. (From *Sound Absorbing Materials*, Evans & Bazley. H.M.S.O.)

155

particle velocity within, say, a 1-in.-thick layer will be low. At higher frequencies, however, that is short wavelengths, the particle velocity within the absorber will be high. Increasing the thickness increases

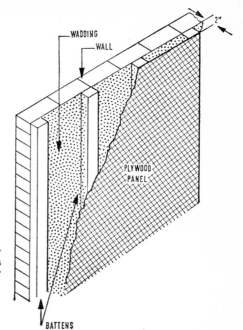

FIGURE 7.13
One type of low frequency absorber uses a non-porous panel fastened over air or wadding-filled cavities.

the particle velocity within the material for the longer wavelengths and thus increases the absorption at the lower frequencies.

Another way to increase the absorption at the low frequencies is to mount the material away from the wall, so that there is an air space behind the material. This is equivalent to increasing the thickness.

Because of the thickness requirements, porous materials are impracticable for low frequency absorption. Below approximately 300 c.p.s., therefore, some other, more effective means of absorption is required.

Resonant Absorbers

Resonant methods are used for low frequency absorption. In the first type, non-porous panels are caused to vibrate by the sound wave and heat is dissipated within the material and at its mountings. The resonant frequency depends on the weight of the panel and the stiffness of the trapped air behind it. The latter in turn depends on the depth of the air cavity. Thus, when absorption is required at a particular frequency, the resonant frequency of the absorber can be adjusted by a suitable choice of the panel weight and air cavity depth. The absorption is often increased by using a layer of porous material behind the panel. Wood and hardboard are two examples of materials used for the panels (see Fig. 7.13).

The second form of resonant absorber makes use of what is called a Helmholtz resonator, where a column of air is made to vibrate. This is rather like the vented loudspeaker cabinet working in reverse (see p. 59). When the sound waves of the appropriate frequency strike the air in the neck, the box resonates and there is considerable air movement in the neck. The air passes through porous materials placed across the neck, which causes heat to be produced due to friction. As with the loudspeaker cabinet, control of resonant frequency is by means of adjusting the volume of the enclosed air and the volume of the neck. Perforated hardboard is now being widely used for acoustic treatment. The rows of holes form an array of Helmholtz resonators, a layer of porous material being placed behind the panel to improve the absorption.

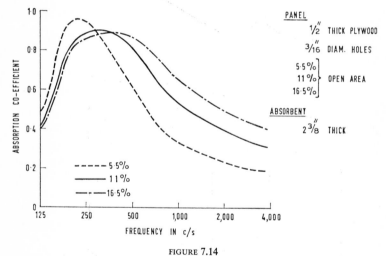

FIGURE 7.14

To obtain low frequency absorption a heavy panel is used and the graph shows how, if this panel is perforated, the open area controls the amount of absorption at high frequencies. (From *Sound Absorbing Materials*, Evans & Bazley. H.M.S.O.)

Resonant methods are normally used at low frequencies and they are most often used to provide general low frequency absorption in large orchestral studios and in concert halls. They are also used in

157

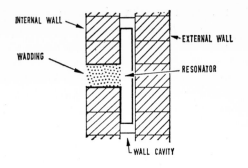

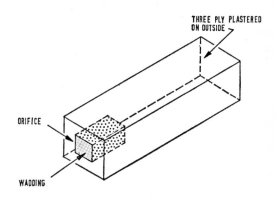

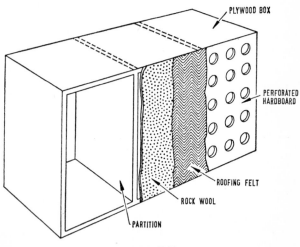

FIGURE 7.15

More elaborate types of absorbers. (*a*) Shows a Helmholtz resonator built into a wall, (*b*) a pre-fabricated Helmholtz absorber, and (*c*) a roofing felt membrane absorber.

small studios to control the room resonances and in factories, etc. With all resonant methods, it is necessary to note that they depend on a body being set into vibration. But a vibrating body can also act as a source of sound, and it is essential to ensure that the energy is dissipated quickly. Otherwise, the so-called absorber may continue to vibrate after the original sound has died away, and radiate sound back to the room. The absorber has what is called a decay time of its own, and this should be less than the reverberation time of the studio.

The absorption of the resonant types is generally quoted in the same manner as that of porous materials, with the absorption co-efficients being quoted at various frequencies.

Air Absorption

In modern acoustic practice, architects and designers can refer to published tables giving the absorbent co-efficient for the wide variety

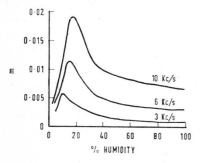

FIGURE 7.16
The effect of the air as an absorbent depends on frequency, humidity, temperature and the volume of the room. Air absorption is given by the expression $4mV$, and the graph shows how the factor 'm' varies with frequency and humidity at 20°C.

of materials and methods used in sound absorption. It is often necessary to include a factor to show how the absorption varies due to the air itself. This absorption is only important above about 4,000 c.p.s.: it increases with frequency, and also depends on humidity and temperature. In the absorption data, this information is usually given the symbol m. The absorption in sq. ft units due to the air is then 4 mV, where V is the volume of the room (see Fig. 7.16).

Total Absorption

If we now consider all the absorption present in a room, due to the structure of the room itself and the materials present, we arrive at a total of sq. ft units which is used in the reverberation time formula. Since the absorption varies with frequency, the reverberation time must be calculated separately at various frequencies. This information, along with the practical experience of the designer, gives predictable results which are generally acceptable to a wide range of listeners.

159

SOUND INSULATION

UP to now we have discussed the control of sound from a source within the studio. The sound from the speaker or artist plus the reverberation should be the only sounds to reach the microphone. But in this day and age it is difficult to ensure that no other sound is picked up. We have to insulate the studio against all possible sources of noise—here defined as unwanted sound.

There has been over the years a gradual rise in the noise level in our everyday surroundings. Traffic along roads and streets has become heavier, and the proximity of airports to urban areas causes considerable annoyance due to the landing and take-off of modern aircraft. There are of course many sources of noise which have always been with us, the clatter of china and of conversation in restaurants, the noise of typewriters in offices, noises due to sinks and toilets, noises in school classrooms, noises from the next door neighbour's hi-fi equipment or television set, and even loud conversation. This is just a sample list of the many annoying sources of noise.

Obtaining adequate sound insulation is a difficult problem and is usually expensive. But, before examining some of the ways it can be done, let us first see how the noise is transmitted.

There are two ways sound can travel from one part of a building to another. There is air-borne sound and impact sound. Sound by air-borne paths can be direct, for example, through an open window or past a badly fitting door. Indirect methods occur when the sound waves from outside cause vibrations in ceilings, windows, walls, etc., so that these vibrate and radiate into the room, or when the sound actually travels through the walls, ceiling, floor, etc.

Impact sounds include footsteps on an uncovered floor, machinery vibrations and slamming doors. The source causes vibrations in the structure itself, and these vibrations can easily travel considerable distances through the materials used in modern buildings, namely steel, concrete and brick.

If possible, of course, steps should be taken to stop the noise from being generated. The impact noise due to footsteps along a corridor

TABLE 8.1. *Maximum Permissible Noise Levels* (*dB above*
0·0002 *dyn/cm²*)

Octave Band (c/s)	Criteria			
	A	B	C	D
37–75	53	54	57	60
75–150	38	43	47	51
150–300	28	35	39	43
300–600	18	28	32	37
600–1,200	12	23	28	32
1,200–2,400	11	20	25	30
2,400–4,800	10	17	22	28
4,800–9,600	22	22	22	27

These criteria refer to noise due to ventilation plant, traffic or similar sources.

Criteria A

Concert halls where best possible conditions are required.

Criteria B

Concert halls where A is not possible
Broadcasting Studios
Opera houses
Theatres (more than 500 seats).

Criteria C

Theatres (up to 500 seats)
Music Rooms
Classrooms
Assembly Halls
Conference Halls (for 50).

Criteria D

Cinemas
Churches
Courtrooms
Conference Rooms (for 20).

From *Acoustics, Noise & Buildings*, by Parkin and Humphreys. Published by Faber & Faber.

can easily be reduced by covering the floor with carpet or linoleum; doors can be fitted with slow closing mechanisms to prevent slamming; proper design of water fittings can reduce considerably the noises set up in a plumbing system. In ventilation systems, the trunking can be an excellent generator and transmission path for both air-borne and impact noise and here careful design and layout are necessary. Later in this section we shall discuss the various acoustic techniques applied to ventilation systems. Since the steel 'core' of a building can be an excellent transmission path for impact noises, it has been omitted in some broadcasting studio centres. The building core is made of a massive structure to retain strength but impede the transmission of sounds from floor to floor.

L

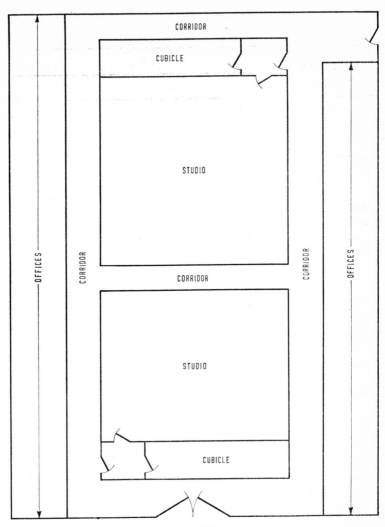

FIGURE 8.1

By correct layout of premises good insulation can be achieved. Broadcast studios, for which good insulation is essential, are protected from street noises by shielding them with corridors and offices. The corridors are treated with soft floor treatment to prevent the generation of impact noise from footsteps and the ceilings are covered with absorbent tiles.

Although a certain amount can be done to prevent noise being generated, by far the biggest problem is to keep out the noises over which we have no control. And here the best form of insulation is isolation. By careful layout of buildings, the areas requiring quiet

162

conditions can be isolated from sources of noise. The first step in the design of broadcasting centres, film studios, recording studios and concert halls is to arrange that the areas which can tolerate somewhat noisy conditions shield those areas in which quiet conditions are imperative (see Fig. 8.1).

It is also important to see that a source of noise is not placed adjacent to a quiet area. In one case quoted in the literature, a gymnasium was placed immediately above a lecture theatre; the effect on the lecturer and his audience can be imagined. Although some relief in situations like these can be provided, the obvious answer is proper layout. Further degrees of isolation can also be contrived to attenuate both air and impact borne noise. A 'box within a box' is made so that the inner surfaces are not in rigid contact with the main structure. This is done by 'floating' the floors on resilient mountings, suspending false ceilings on resilient hangers, and so on. So much for the general problem. Now we shall examine the problem in more detail and quote some typical methods and treatments.

Transmission Co-efficient and Sound Reduction Factor

The transmission co-efficient of a partition is the fraction of incident sound energy which is transmitted.

The efficiency of a sound insulating partition is given by the sound reduction factor, which is defined as follows:

Sound Reduction Factor

$$= 10 \log_{10} \left(\frac{1}{\text{Transmission Co-efficient}} \right) \text{decibels}$$

In practice, an average sound reduction factor is usually quoted. This is measured over a frequency range from 100 c/s to 3,200 c/s. However, most insulating materials and methods have factors which are frequency dependent. Roughly speaking, the insulation increases 5 dB per octave and so, when an accurate assessment is required, the sound reduction factors at all frequencies must be quoted.

Air-borne Noise and its Reduction

Walls. The sound reduction factor of a partition, as far as air-borne noise is concerned, depends on the porosity of the partition material and the degree of vibration induced in the partition by the incidence of the sound wave.

The reduction factor for an unplastered wall is very low, a representative figure being 15 dB. When the wall is plastered, the factor rises to about 40 dB. This increase of 25 dB shows how important it

TABLE 8.2. *Octave Analyses of some Common Noises*

Noise	Distance (ft)	Sound-pressure level (Octave Bands)								Remarks
		37–75	75–100	150–300	300–600	600–1,200	1,200–2,400	2,400–4,800	4,800–9,600	
Large (4 engines) jet air-liner	125	112	121	123	124	123	120	117	109	Maximum values when passing overhead at take-off power. No mufflers
Single-engined jet fighter	125	102	114	116	116	117	115	111	102	Maximum values when passing overhead at take-off power
Large (4 engines) piston-engined air-liner	125	111	117	114	108	107	108	106	97	Maximum values when passing overhead at take-off power
Electric trains over steel bridge	20	94	93	99	99	95	84	81	73	
Curb-side, main road in London at rush hour	15 (average)	78	81	81	79	72	67	63	55	
Electric trains	100	77	77	76	74	73	67	59	54	In open air
Pneumatic drills	125	75	72	72	66	69	71	67	65	
Riveting on large (20 ft by 15 ft) steel plate	6	88	96	105	106	111	109	113	110	
Nylon Factory	Reverberant sound	87	86	92	93	97	97	96	87	Up-twisting process
Weaving shed	Reverberant sound	78	71	77	81	86	86	84	78	
Canteen (hard ceiling)	Reverberant sound	52	54	59	67	67	61	55	49	Average levels. Peak values up to 20 dB higher
Typing office with acoustic ceiling	Reverberant sound	68	64	60	56	55	55	53	50	Ten typewriters, one teletype machine
Male speech	3	52	55	59	66	65	60	52	40	Average values

From *Acoustics, Noise and Buildings*, by Parkin and Humphreys. Published by Faber & Faber.

is to reduce the porosity of any partition if air-borne noise is to be attenuated.

The vibration of a partition depends on several properties but the important ones are the mass, stiffness, elasticity and density of the material. In practice, mass is the most important property in indicating the sound reduction factor to be expected since, if the mass or weight is increased, the factor also increases. For example, a wall of 4½-in. brick weighs about 55 lb/sq. ft and gives a sound reduction

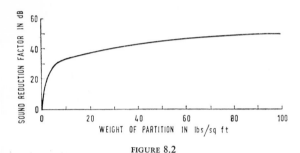

FIGURE 8.2

The sound reduction factor of a partition depends on the weight per unit area, as shown by the graph.

factor of 45 dB. If the thickness is increased to 9 in., the weight is doubled and the reduction factor goes up to about 50 dB. By comparison, clinker block 2 in. thick has a weight of 18 lb/sq. ft and gives a factor of 30 dB (see Fig. 8.2).

The figures for the 4½-in. and 9-in. brick walls show that, if very good insulation is required and weight were the only criterion, very massive walls would be needed since doubling the weight only increases the reduction factor by 5 dB. In some cases reductions of over 50 dB are necessary, and this would call for impractically heavy and costly structures (see Fig. 8.3).

Fortunately, by adopting discontinuous methods, good reduction factors with light structures can be achieved. One good example is the technique much used in modern house construction whereby sound and heat insulation are achieved at the same time by building two single brick walls with an intervening air cavity. The sound insulation achieved will depend on how much of the sound vibration or transmission through the first wall is passed to the second. The air cavity is a means of coupling and this can be reduced by increasing the distance between the walls or by hanging a sound absorbent blanket in the cavity. Another means of coupling is the mechanical

165

linkage between the two walls due to the steel ties which are required to strengthen the structure. Special 'butterfly' ties are available which reduce this coupling. Another source of mechanical coupling is the mortar dropped into the cavity during building.

It is also important to note that the insulation, despite precautions, could fall to a low figure if there were mechanical coupling between the two walls through the structure into which the walls are built. This is an example of what is termed a 'flanking path' and it is important to realise that, if there is a low insulation flanking path, it is

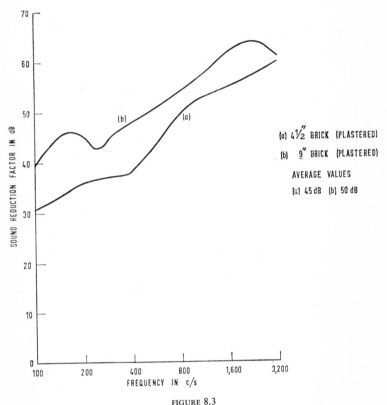

(a) 4½″ BRICK (PLASTERED)

(b) 9″ BRICK (PLASTERED)

AVERAGE VALUES
(a) 45 dB   (b) 50 dB

FIGURE 8.3
The sound reduction factor of brick walls is seen to rise only slightly if a double thickness construction is used.

not much good going to the trouble of obtaining a high reduction factor for the direct path. By using some soft material at the edges of the walls, the coupling between them due to the structure can be

166

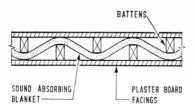

BATTENS

SOUND ABSORBING
BLANKET

PLASTER BOARD
FACINGS

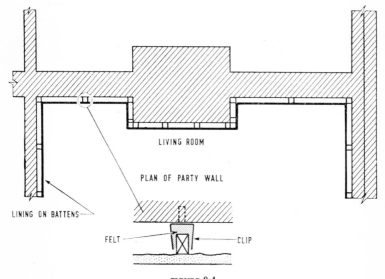

LIVING ROOM

PLAN OF PARTY WALL

LINING ON BATTENS

FELT

CLIP

FIGURE 8.4

The top diagram shows constructional details of a staggered stud partition. The lower drawing illustrates the use of insulation board, retained by felt-lined clips, to improve the sound insulation between adjacent houses. (By permission of H.M.S.O.)

considerably reduced. In practice, however, it is difficult to use soft material as the walls usually have to have good mechanical stability.

Cavity walls with glasswool blankets in the cavity can give sound reduction factors of over 80 dB. This discontinuous principle can be used with lighter building materials, such as clinker blocks or staggered stud partitions with plaster-board facing, and factors in the region of 50 dB can be obtained. The stud partition has a weight of about 10 lb/sq. ft. When we recall that a homogeneous structure of 9-in. brick gives only 50 dB, we realise that the discontinuous method of sound insulation is much more efficient than a continuous material, however heavy (see Fig. 8.4).

Windows. Glass of the type used in windows gives a reduction factor of about 20 dB, while plate glass is some 5 dB better. These

167

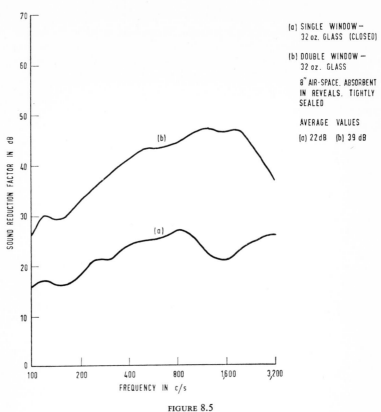

(a) SINGLE WINDOW —
32 oz. GLASS (CLOSED)

(b) DOUBLE WINDOW —
32 oz. GLASS

8" AIR-SPACE. ABSORBENT
IN REVEALS. TIGHTLY
SEALED

AVERAGE VALUES

(a) 22dB   (b) 39 dB

FIGURE 8.5

The sound reduction factors for two types of windows, showing the considerable improvement from double glazing.

figures show how the windows let into a wall are a very weak link as far as insulation is concerned. In a composite structure there is no serious error in assuming that the insulation of the whole is simply the insulation of the weakest member. For a window of 32-oz glass, for example, the insulation when it is closed is about 22 dB. This would be the effective attenuation for the whole wall, and so improvement here is essential if an adequate reduction factor is to be reached (see Fig. 8.5).

Unfortunately, windows are associated with the question of ventilation and in many domestic cases the insulation is sacrificed so that fresh air—with the accompanying outside noises—can enter the room. But where the ventilation question can be tackled by other means, the insulation of the window can be raised considerably by applying the discontinuous principle again and using double-glazed windows. As with cavity walls the reduction factor rises as the spacing between the sheets increases. At an 8-in. spacing, a double window of 32-oz glass can provide insulation up to about 40 dB. The glass sheets should be structurally insulated from each other by building one into a resilient mounting. Further improvement can be obtained by lining the edges of the air-space with porous absorbent material and ensuring that the windows fit tightly into their frames so that there are no air leakage paths (see Fig. 8.6).

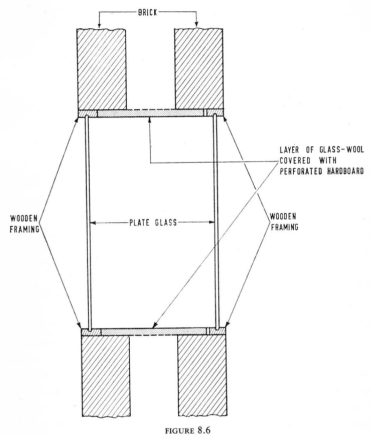

FIGURE 8.6

A typical treatment for double-glazed windows to achieve good sound insulation. The glass is set in resilient mountings. If one of the panes is set on an opening frame, this is lined to give an air-tight seal.

The same techniques are applied to glass partitions. One important problem in broadcasting studios is the glass partition between a studio and its control cubicle. Some authorities apply the technique outlined above using plate glass. Others use triple-glazed windows.

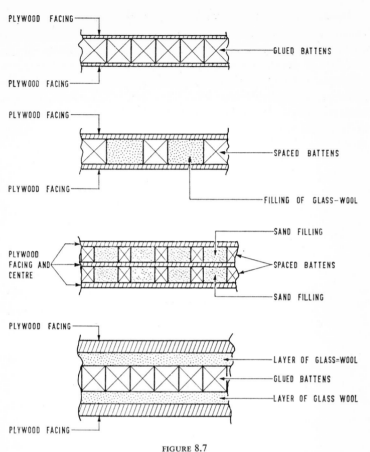

FIGURE 8.7

Some typical door treatments. The first two are of light construction and would normally be used with a sound lock. The second two are designed to give insulation in a single door.

Doors. Doors present the same problems as windows, as far as sound insulation is concerned. They are both necessary but are weak links. The sound reduction factor of a door is almost entirely determined by its weight. For a normal panelled door a typical factor would be 19 dB, and for a blockboard door the factor rises by 4 dB. In many cases, figures well above these are required and in broadcasting centres, hospitals, etc., use is made of heavy doors made up of alternate layers of wood and glass wool. Two important points to watch are the tightness of the door in its jamb and any leakage between the jamb and the wall. Sound absorbent material is used to ensure that these leakage paths are well attenuated.

170

Double doors are often used in the same way as double-glazed windows. These can be of lighter construction and, with a sound lock, give good insulation without the heavy construction required for a single door (see Fig. 8.7).

Floors and Ceilings. An average value of sound reduction for a floor, consisting of boarding on joists with a plaster-board ceiling underneath is 35 dB. A lath and plaster ceiling of the type which used to be common is a better insulator and the reduction factor is raised to 40 dB.

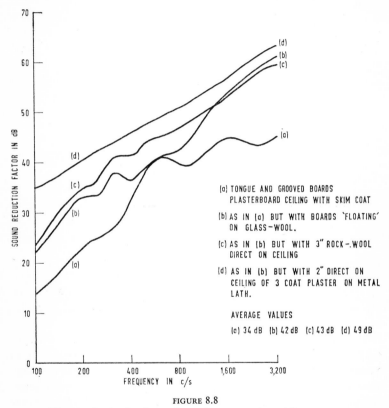

(a) TONGUE AND GROOVED BOARDS PLASTERBOARD CEILING WITH SKIM COAT

(b) AS IN (a) BUT WITH BOARDS 'FLOATING' ON GLASS-WOOL.

(c) AS IN (b) BUT WITH 3" ROCK-.WOOL DIRECT ON CEILING

(d) AS IN (b) BUT WITH 2" DIRECT ON CEILING OF 3 COAT PLASTER ON METAL LATH.

AVERAGE VALUES

(a) 34 dB   (b) 42 dB   (c) 43 dB   (d) 49 dB

FIGURE 8.8

The sound reduction factors for various types of wood floor treatments. All are seen to rise with frequency.

The insulation can be increased by pugging the ceiling to increase the weight. This is done by using a filling on the ceiling boards, the heavier the better, but there are obvious limitations as to the weight-bearing properties of the ceiling. A pugging of 20 lb/sq. ft increases the insulation by 5 dB. This would be provided by a 2-in. layer of dry

171

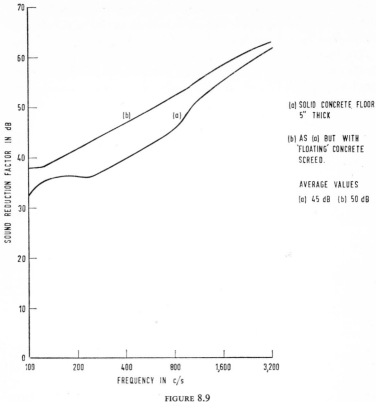

FIGURE 8.9
The sound reduction factors for two types of concrete floor.

sand, and this sort of layer would normally be used on a ceiling of plaster on metal laths or a lath and plaster ceiling. A lighter pugging of a mineral wool can be used with a plasterboard ceiling. A 3-in. thickness is typical, the weight being about a tenth of that of sand so that the insulation is not increased so much as with sand. Where very good insulation is required, such as in broadcasting studios, heavy ceilings are used. These are made of heavy metal lath and plaster supported on light metal beams. The weight of such a ceiling is not less than 5 lb/sq. ft and the deeper the space between it and the structural floor above the better the insulation.

On a layer of resilient material, such as glass-wool or mineral wool spread over the joists, the insulation can be improved compared with the straightforward system of nailing the floorboards direct to the joists. It is essential that there should be no contact between the floor and the joists. If the floorboards are to be nailed, they should be

172

nailed to fresh battens laid on the glass-wool and not nailed through the glass-wool to the joists. The improvement gained by floating the floorboards in such a manner is around 6–8 dB (see Fig. 8.8).

The reduction in sound transmission by concrete floors depends firstly on the weight; a 5 in. layer of concrete, for example, has an insulation factor of 45 dB. As with timber floors, this is improved by

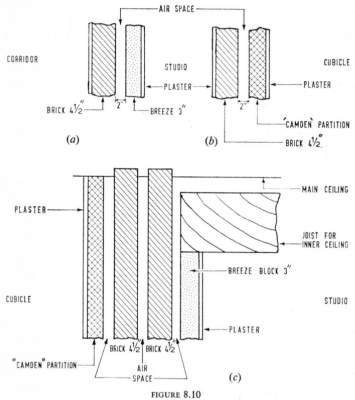

FIGURE 8.10

Construction methods used in broadcasting centres for partitions and inner ceilings in talks studios: (a) between studio and corridor, (b) between studio and own cubicle, (c) between studio and another cubicle. (By courtesy of the BBC.)

'floating' the floor screed on a resilient layer of glass-wool or similar material.

Impact Sound

We have already seen how impact noise is produced and transmitted. To complete the discussion we will examine how it is measured and how it is attenuated in practice.

173

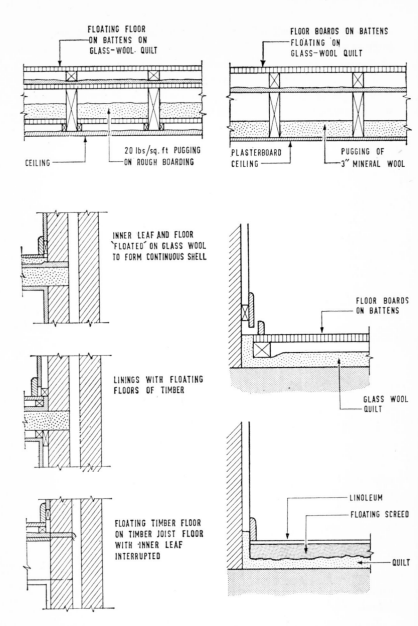

FIGURE 8.11

The upper two sketches show how floors, either wood or concrete, are 'floated' away from the main structure on glass wool quilts. The three left-hand diagrams illustrate insulation details for flatted houses, and those on the right floors on a concrete sub-floor. (By permission of H.M.S.O.)

174

Most of the trouble from impact sound is due to floors and it is difficult to measure the effect of typical blows—footsteps, scraping of chairs and so on. To overcome some of these difficulties a standardised procedure has been adopted to provide comparative data between different treatments. Blows are produced by a 'tapping' machine which simulate the effect of footsteps. The sound pressure in the room below the floor is measured and referred to the acoustic

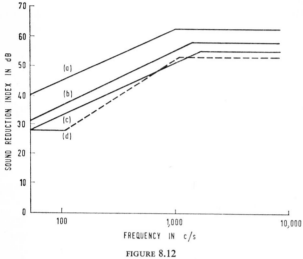

FIGURE 8.12

In a broadcasting centre, where several studios are grouped together, the degree of insulation required between various areas is as shown: (a) studio from cubicle with another programme, (b) studio from studio, (c) cubicle from cubicle, (d) studio from own cubicle. (By courtesy of the BBC.)

reference level 0·0002 dyne/sq. cm. The noise produced is analysed in octave bands over a range usually from 50–1,600 c/s or from 100–3,200 c/s; the transmission is frequently dependent but an average figure is normally quoted. The acoustical conditions of the receiving room will affect the pressure readings and, to allow for this, readings are taken at several points in the room and the results averaged. In addition the absorption present, or the reverberation time, is taken into account and referred to standard values, 100 sabins or 0·5 seconds.

To illustrate the effectiveness of some typical treatments, we can start with a bare concrete floor. Using the tapping machine, the level produced below this floor will lie somewhere in the range 70–75 dB above 0·0002 dyne/sq. cm. A covering of linoleum will reduce this by 3 dB, and a reduction of 20 dB is produced by a thick (⅜ in.)

175

carpet. Another way of specifying the reduction is to use phons. The pressure levels above 0·0002 dyne/sq. cm are referred to contours of equal loudness for constant spectrum noise and loudness of the received noise can then be stated in an equivalent number of phons.

Some authorities specify a bare concrete or timber floor as giving 0 phons insulation and then quote the improvement due to treatment as so many phons. With concrete a layer of $\frac{1}{16}$ in. sheet rubber on $\frac{1}{4}$-in. sponge rubber gives a reduction of 20 phons. This emphasises the comment at the beginning of this section that if possible the noise should be tackled at the source. Covering the concrete with soft material prevents a heavy impact noise being created.

As with air-borne sound, considerable improvement is obtained by using discontinuous methods of construction. Floating a 2-in. thick concrete screed on a quilt of eelgrass gives about 17 phons reduction. If floorboards are used, the battens should be isolated from the concrete by some resilient material. A $\frac{1}{2}$-in. thick quilt of mineral wool will give a reduction of about 15 phons.

With timber floors a carpet on underfelt gives 10 phons reduction and approximately the same is obtained by floating the floorboards on quilts.

Ventilation Systems

Where good insulation is essential, the room must be sealed off from the various paths of transmission and one of the most important is the direct one connecting to the outer air. This makes the installation of some ventilation plant imperative. The plant must be well designed and installed, otherwise the system can become a noise source and a good link for air-borne and structure transmitted sound.

Where a ventilation system is used in places such as concert halls and broadcasting studios, it is an obvious requirement that it should operate quietly. A potential source of noise is the intake fan. This can generate air-borne noise from both the motor and the fan as well as structural-borne noise due to vibration. The latter is guarded against this by mounting the machinery on resilient supports so that it is isolated to a great extent from the building. Where metal ducting is used, it is also common to join the duct to the intake fan with a canvas coupling so as to avoid vibrations passing down the duct walls. To prevent the metal duct being a generator and a transmission path for impact sound, the duct is sometimes supported on resilient hangers and the metal cut at several intervals along the trunking, the cuts being bridged over with canvas couplings.

The fan is the main source of air-borne noise, the amount and frequency content depending on the fan speed. A high-speed fan generates a great deal of high frequency noise; a low-speed one produces lower levels of noise but it is of low frequency. Because it is easier to absorb high frequencies than low ones, a small high-speed fan is preferable to a large slow-speed one. But of course the choice depends on circumstances.

Another source of noise is the air itself and, if the air flow velocity is high, the air, in passing along the ducts, over obstructions and through grilles, can produce considerable noise. As a rough guide, the maximum velocity should be about 600 ft/minute. The noise produced by the air flow is due to turbulences and eddy currents being set up. To avoid these, there should be no obstructions in the duct, and changes of cross-sectional area and direction should be gradual and not abrupt.

Absorption material is introduced into the ducts either by lining the duct walls or by having splitters in the air path. These are made of an impervious core covered with absorbent material.

DOMESTIC ACOUSTICS

It is generally impossible to carry out in the home the elaborate procedure used in professional work but considerable improvements in acoustic conditions can be obtained by modified, simple means. In this section we examine these means and show how they can be applied, firstly for good reproduction—both monaural and stereo— and then to convert a room into a studio suitable for recording.

When looking at the problem of reproduction it is perhaps pertinent to enquire what the end product is. Purists say that the aim should be to recreate in the home the acoustic conditions which existed at the source, be it a broadcasting or recording studio. But this is not possible. On the one hand the transmission network has limited dynamic range and, on the other, the acoustics of the listening room modify the result. There are other factors as well but the final conclusion must be that the aim of any system is simply to produce an acoustic environment which the listener will find acceptable under the rather unnatural conditions in which he is listening. This means that in the end one must judge the final reproduced sound on one's own taste, backed up, of course, by past experience.

The unnatural conditions mentioned above apply to both mono and stereo but much more so to mono. Although much skill goes into preserving the light and shade of the programme material there is no doubt that a single channel inhibits the production of acceptable listening conditions. One is more conscious of the defects of the system, and the results have been described as 'listening through a hole in the wall'. With stereo the results are better because more 'information' is available in the listening room, and this gives a spaciousness to the reproduced sound which the monaural system lacks. However, a mono system is comparatively simple and gives acceptable results for many purposes.

Perhaps a suitable starting point in discussing reproduction systems is to examine the sound level which is required to give acceptable results. Tests have been carried out on monaural systems to measure the preferred maximum sound level (in dB above 10^{-16} watts/sq. cm) for various types of programme. They were carried out by Somerville and Brownlees* of the BBC in 1949 using music and speech as material for listening groups made up of members of the public, musicians and engineers. For symphonic music the general public preferred 78 dB, the musicians and engineers 88 dB. For speech the respective figures were 71, 74 and 80. These tests were carried out under conditions similar to those in the home; the principle was to allow the listeners to adjust the loudness of the programme till they were satisfied and then measure the pressure 18 inches from the head by a weighted sound level meter. Much additional information was derived about the effect of age on preferred levels but perhaps the most significant point is that the maximum levels are more than those encountered on listening directly to the sources of programme.

TABLE 9.1 *Preferred Listening Levels*

Programme	Public		Musicians	Programme Engineers		Engineers
	men	women		men	women	
Symphonic music	78	78	80	90	87	88
Light music	75	74	79	89	84	84
Dance music	75	73	79	89	83	84
Speech	71	71	74	84	77	80

Other investigators have carried out similar tests on preferred frequency ranges and established some most significant points about the effect of monaural reproduction. The choice of a restricted bandwidth by many listeners of different types and the effect of distortion and noise on bandwidth choice produced interesting and sometimes surprising results.

The effect on these results of changing to stereophonic reproduction is interesting, tests having shown that a stereophonic system of very restricted bandwidth was considered equal to a monaural system of more than four times the bandwidth. For more detailed information, the reader is referred to Chapters 4 and 6 of *High Quality Sound Reproduction* by Moir, published by Chapman and Hall.

However, despite some of these surprising conclusions about various listeners' preferences, the hi-fi enthusiast will continue to insist on a standard of reproduction which requires a system of adequate power and wide frequency range.

Placing of Speakers
Assuming that the reproduction equipment is of good quality, the next point to consider is the room itself and how it affects the

* Listeners' Sound-Level Preferences—Somerville & Brownlees, *BBC Quarterly*, Jan. 1949.

reproduction. The room acts as an acoustic extension of the loudspeaker and its acoustic properties can introduce colorations to the final reproduction.

A major problem in reproduction is to obtain smooth and extensive bass response. A fundamental difficulty is to obtain sufficient air loading on the loudspeaker so that there is adequate transfer of energy at low frequencies. The corners of a room are points of high acoustic impedance and the coupling of a loudspeaker placed in a corner to the air in the room is such that the bass response is considerably enhanced. Corner enclosures have been popular for many years for this reason. However, one point has to be borne in mind, that all the room resonances can be excited from the corners and an isolated resonance may produce audible colorations.

The frequency of any such coloration would depend, as we have already seen, on the dimensions of the room and, although studios are often designed with dimensions which control the distribution of the modes, the spacing of room modes is not normally a consideration when a house is being built.

Therefore, if the bass response does not sound 'clean' there is one solution worth trying, to move the loudspeaker, if possible. Moving the speaker down one wall away from the corner will affect the production of the room resonances and may well attenuate any troublesome coloration. But an improvement at one frequency may be offset by the isolated stimulation of a fresh resonance.

From this it can be appreciated that the position of the loudspeaker is open to experiment and there is no unique correct position. As mentioned earlier, the aim is to produce a result which one likes, so it is certainly worth trying several loudspeaker positions to see if there is one which provides an acceptable compromise.

Another point often raised in connection with monaural reproduction is whether more than one speaker should be used. This is really a question of how the energy is distributed—over what area and at what frequencies is the sound distribution made uniform. At high frequencies there is no doubt that there is a tendency for the source to appear small and clearly identifiable. A less directional response is sometimes asked for and many systems have been suggested to give a more diffuse sound and make the source appear less localised.

Some systems have attempted to provide omni-directional characteristics by using several low and high frequency units radiating in different directions. The effect is claimed to give a spatial distribution to monaural sound which approximates to sterephony. A simple way of trying the effect of an omni-directional loudspeaker is to play an ordinary loudspeaker into the corner of a room. This scatters the sound and simulates omni-directional distribution to a certain extent.

It is important to note, however, that not everyone is agreed that an omni-directional distribution is necessarily the ideal for all types of programme. Speech, for example, is not always acceptable when heard over an omni-directional system. Experimental subjective work tends to indicate that the preferred polar distribution should be that of the original sound source.

All of the foregoing suggests that, as with placing a single loud-speaker, experiment is the answer; using two speakers and altering their positions until one is satisfied. Of course it may well be that for listening to certain programme material it will be better to switch off one of the units.

With stereo it is of course essential to have at least two loudspeakers. Although various suggestions have been made for using a central loudspeaker common to both channels for the low frequencies, this was deprecated and most systems now use two electrically separate units. The word 'electrically' is used here since the speakers in several commercial units are mounted in the same cabinet either facing forward or radiating from the ends. By reflection from the walls, it is claimed that the width of the sound stage is made larger than the physical separation of the loudspeakers. This is true to a certain extent but it is better for the loudspeakers to be physically separate since this allows much more control of their placing.

As with monaural reproduction there are many opinions as to the best directional characteristics for stereo loudspeakers. Some authorities recommend omni-directional units, some cardioid. What is certain is that the loudspeakers should have similar characteristics and that their positioning depends on their directional characteristics.

To start with, you should try the commonly recommended arrangement in which the loudspeakers and the listener form an equilateral triangle, the loudspeakers being angled towards the listener. This gives a good stereophonic effect but the area over which this is available is restricted and the effect is diluted towards the back. By altering the angle of the loudspeakers, the position of this area of good stereophonic effect can be placed so as to suit the seating arrangements. Another suggested method is to make the axes of the radiations from the loudspeakers cross at a point in front of the seating area.

So much for the relative positions of the loudspeakers and the listeners; now we discuss the positions of the loudspeakers in the room. It is better for the loudspeakers to radiate down the room with their axes crossed at some convenient point depending on the loudspeakers and the properties of the room. It may sometimes be thought convenient, where the fireplace is on a long wall, to place the loudspeakers on either side of the fireplace so that they radiate across the room. However, Moir comments that such an arrangement has never been found to give good results.

Thus, if possible, start with the two loudspeakers in the corners, radiating down the room with their axes crossing half-way down.

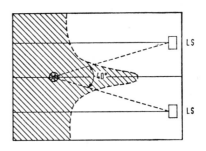

FIGURE 9.1
The shaded area indicates where satisfactory stereo reproduction results, based on a subtended angle of 40°. (From *Loudspeakers*, by E. J. Jordan.)

This will form a basis for experiment in adjusting the positions to get optimum results.

Listening Room Acoustics and Treatment

The possible effect of the listening room acoustics has been mentioned. In this section we shall discuss how the room can be improved by simple acoustic treatment both from the point of view of improving the reproduction and keeping out extraneous noises. At the same time it is of course important to bear in mind that one's 'hi-fi' equipment may be regarded as a noise by the next door neighbour.

With normal house construction there is usually absorption of the bass due to vibration of the structure, the middle and high frequencies being absorbed by the furnishings—carpets, soft furniture and curtains.

The usual domestic arrangement with carpet on underfelt concentrates a large amount of the absorption on the floor and leaves the other surfaces reflecting a great deal of sound. Of course the curtains at the windows will absorb some sound, but the extent depends on the material and how long the curtains are. Heavy thick ones hanging right to the floor will provide good absorption. A better balance of absorption could be obtained by reducing the amount of carpet on the floor and hanging lengths of thick curtain material or drapes on the walls. These need not cover the entire wall surface, but the lengths should be distributed so that each wall has some absorbent material. By arranging the material in thick folds, the absorption can be made to spread down to the lower frequencies. The ceiling too should receive some attention. Lightweight acoustic tiles, either fibrous or plastic, make this an easy task. These can also be used on the walls instead of drapes and, arranged in random patches surrounded by wood framing, can prove acceptable as far as room decoration is concerned. The aim of such treatment is to control the standing waves which can be set up between the surfaces of the room.

With stereo the problem of reflections is much more serious since they can cause confusion and degrade the stereo effect. Usual domestic arrangements do not provide enough distributed absorption, as we have seen, so it may be necessary to introduce absorbent materials. A treatment along the lines outlined above should ensure that the spatial distribution of sources on the sound stage is retained.

Noise reduction is important since it is most annoying, when listening to music, to be distracted by the noise of traffic, the banging of doors or such like. Earlier in the chapter we have examined the problem of sound insulation and seen some means of providing control over noise transmission. It was mentioned that good sound insulation is usually expensive, but there is perhaps one point that ought to be borne in mind and that is that many forms of sound insulation simultaneously provide heat insulation. A ready example is the double-glazing of windows; not only will this give much improved sound insulation, but it will also prevent a considerable amount of heat from going to waste.

Remembering that isolation is the best form of insulation, it is a good idea to have the listening room as far away as possible from sources of noise. An example is in semi-detached houses where the entrances and halls are on the party wall. This isolates the main rooms of the two houses so that they do not interfere. In this case double-glazed windows would be a first step towards protection against street noises. With downstairs rooms, the door of the room may be near the outside door and, unless precautions are taken, this could be an easy path for outside noises. Ensuring that the outside door is well sealed in its frame either by plastic, foam rubber or metallic strip is essential. Of course adding a porch complete with door, or fitting a second door to form a sound lock if space permits, will make a very considerable improvement.

Unfortunately there are many semi-detached or terraced homes where the main rooms are on the party wall and this can pose a severe problem in sound insulation. One step towards a partial solution is to cover the wall with insulation board. This should be attached by felt-lined clips to battens nailed to the wall. Another method is to form a framework of thick timber about six inches away from the party wall and cover this with insulating board, wood wool slabs or similar material. Where this partition butts on to the floor, ceiling and walls should be well sealed to avoid direct air paths. It should be borne in mind that flanking paths may exist, and it may well be that treatment of the party wall is insufficient to provide as much insulation as is required. There must obviously be a compromise between cost and requirements.

With flats the problem can be even worse than in houses, as transmission in the vertical direction has to be considered as well as transmission in the horizontal. The various treatments suggested for ceilings and floors provide means of improvement.

Providing a Studio

A very considerable improvement in the quality of home recordings is possible if some attention is paid to the room acoustics and sound insulation.

We should consider insulation first, as extraneous noise is often the first obvious fault on a recording. In the previous section this problem has been discussed and some solutions outlined. The actual solution used depends on the particular situation but it is essential to provide adequate insulation since, if it is not, noise can ruin a great deal of careful work.

Assuming that noise level can be kept to a low level, what treatments can be used to control the reverberation? For speech, one immediate point to be made is that, although a room may be satisfactory to listen in, it may not make a successful studio. Our binaural hearing mechanism enables us to reject a great deal of reverberant energy, but a microphone cannot do this. It is therefore necessary to control the reverberation so that an adequate ratio of direct to reflected sound is obtained. The reverberation time for a room suitable for speech should be quite short; broadcast talks studios are designed to have reverberation times of around 0·3 sec.

In the home, simple treatments of a porous material, such as mineral wool, or fibreglass, can be used as a covering over most of the wall surfaces. This will provide considerable absorption. One way of mounting it is to screw battens to the walls, of say 1 inch thick, forming boxes 4 feet wide. The porous material can then be used as a filling, being retained by wire netting or something similar. Porous materials available for such treatment are not exactly pleasant to look at, so hang a curtain in front to obscure the material and make the 'decor' more acceptable. Of course the curtain must not acoustically obscure the material. It should be of open weave material so that the sound waves can easily pass through.

We have already seen how the absorption of a porous material depends on its thickness and on any covering it may have, e.g. perforated or slotted hardboard or perforated metal. If it is possible to

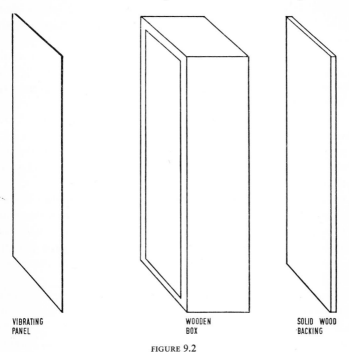

VIBRATING PANEL   WOODEN BOX   SOLID WOOD BACKING

FIGURE 9.2

The essentials of the resonant panel type of bass absorber. The box is of solid timber and, with the back and panel, forms an air-tight box. The panel can be of plywood, hardboard or roofing felt. The absorption is improved by fixing inside the box a layer of porous material such as rock-wool or glass-wool. The resonant frequency is governed by the volume of the box and the panel weight.

increase the thickness of material used, the absorption will spread down to a lower frequency. Alternatively, having an air space behind the material will do the same thing; 3 inches would be suitable.

At high frequencies there is a tendency to have too much absorption. This can be corrected by covering some of the porous material with perforated or slotted hardboard.

185

To sum up the treatment of walls for speech recording, some of the area, say two-thirds, should have a layer of porous material covered by curtains, with the remainder covered by hardboard. The treatments should be intermixed; that is on each wall there would be some curtaining and some hardboard.

The floor should be covered with a good carpet on underfelt. The ceiling should be treated with a quantity of the acoustic tiles sold for this purpose.

With small rooms there is always the problem of boominess due to room resonances. Treatment with porous materials will not have much effect, and putting bass cut in the microphone circuit will improve matters as far as the recording is concerned. A directional microphone such as a figure-of-eight or a cardioid will discriminate against some of the room resonances. The position of the microphone is important, so it is worth trying several positions and recording each to find which gives the best results.

When using a room for recording a talk, one should remember that objects such as lampshades or vases may vibrate when someone is speaking close to them and produce coloration. This will be revealed in the form of 'ringing' noises.

If it is decided to go further in acoustic treatment and provide some bass absorption to control the room resonances, panel absorbers are the answer. These can be made by forming on the wall a box framing of wood and covering it with plywood, hardboard or roofing felt. The depth of the box will control the resonant frequency and by having several boxes of different depths, say 3 inches, 4 inches, 5 inches and 6 inches, there will be a spread of resonant frequencies to give absorption over a range of bass frequencies. The size can be 3 feet × 2 feet. As was suggested for the erecting of porous absorbers, each wall should carry some of the units (see Fig. 9.2).

For recording a play, if the room is big enough it should be given a 'live' end and a 'dead' end. The latter should be treated with plenty of porous material, the live end should only be partially treated, so that the reverberation time is longer. A heavy curtain can be used to separate the sections, and the two different areas will provide a change of acoustic which is often useful for changing from one scene to another.

Acoustic screens are useful on many occasions when recording. They may be made absorbent on one side and reflecting on the other. A cheap and simple way to make these is to construct a wooden framework of wood battening, say 2 inches thick. Hardboard, with its polished surface outwards, should be nailed over the frame. Then the box should be filled with glass-wool and covered with some fabric of open weave. With screens it is important to remember that their effectiveness depends on size; the bigger they are the lower the frequency at which their screening properties become appreciable (see Fig. 9.3).

Music, it must be remembered, normally requires a longer reverberation time than speech. This means that there should be less absorption. If one is recording in an ordinary living room, where the

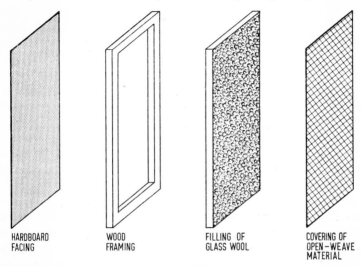

HARDBOARD   WOOD   FILLING OF   COVERING OF
FACING   FRAMING   GLASS WOOL   OPEN-WEAVE
                                                        MATERIAL

FIGURE 9.3
A cheap acoustic screen can be constructed using a wood framing of 2 in. × 2 in. timber. Together with a hardboard facing, the framing forms a box which is filled with glass-wool. It is wise to nail some wire netting to the framing so that the glass-wool is retained in position when the screen is standing up. To finish off, a covering of open weave material should be placed over the wire netting and tacked down on the framing. A good size for the framing would be 5 ft × 3 ft.

carpet is a heavy absorber, a first step would be to reduce the area of carpet in use so that some reflections can be obtained from the floor. As with speech, test recordings should be made of different positions and conditions to see if there is one which is better than the others.

If the room is large, of course, the reverberation time may be too long for some types of music so that additional absorption may be required. Some drapes of heavy curtains at the walls will improve matters. There are so many possible circumstances with musical recordings that it is impossible to go into details for every case. But if one remembers the general principles, the answer lies in experimenting until the desired result is achieved.

ACOUSTICS IN TELEVISION AND PUBLIC ADDRESS

So far we have been discussing acoustic problems in concert halls, studios and rooms in which good sound is the only criterion. Their design, acoustic treatment, placement of microphones have one common aim—to produce a pleasant sound. But a very important problem in modern times is how to produce good sound in television studios. The conditions are very different from those in sound broadcasting studios. First of all, a great deal of technical equipment is present in the studio, cameras, lighting apparatus, scenery, etc. Then there are members of the technical crew, make-up and wardrobe, lighting electricians and other staff on the floor of the studio in addition to the artistes.

Special Problems of Television Studios

To accommodate all the necessary equipment, the studios have to be very large, having volumes of several hundred thousand cubic feet. A typical volume would be about 350,000 cubic ft. This large studio, with all the equipment, costs a great deal of money and consequently there are fewer television studios than there are sound studios. Whereas the latter can be designed for specific types of programme— speech, music or light entertainment—television studios have to be used for all types of programmes, and this leads to the acoustics being a compromise. In practice, this means that the reverberant qualities of the studio have to be made more suitable for speech than for music. Microphone placement is difficult in lively acoustics, when a distant microphone technique is required to keep the microphone out of shot. Thus the treatment of the studio aims at reducing the reverberation

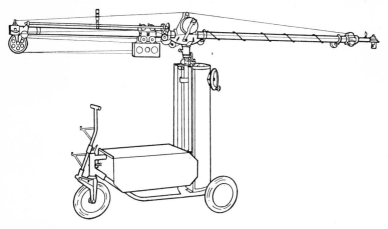

FIGURE 10.1

A dolly-mounted microphone boom of a type used in film and television studios to allow the operator to follow actors from point to point round the studio.

time, so that the acoustics are fairly dead in spite of the relatively large volumes.

Some idea of the extent to which the reverberation time is reduced can be got by considering two studios of the same volume, one designed for sound and one for television. The former would be used for music broadcasts and have a reverberation time of about 1·8 sec. The reverberation time of the television studio would be around 1·0 sec. at the most.

With such low reverberation times, this means that when music programmes are broadcast on the television service, special techniques and equipments have to be used to create the illusion that the orchestra is playing in reasonably good acoustic conditions.

A fuller description of these techniques will be given later, but for the moment we will consider how the microphones are used when speech is being broadcast.

Boom Microphones

If a play is being televised, for example, it is essential that the microphone should stay out of shot and yet be near enough to obtain a

189

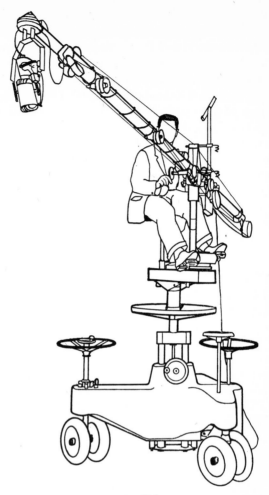

FIGURE 10.2

A second type of microphone boom. A weight box gives variable
counter-balance as the telescopic arm is extended, giving the operator
a wide measure of control.

good ratio of direct to reflected sound. With camera changes, both
in position and in the angle of view of the lens being used, and the
movement of the actors, the microphone must be mobile. Static
microphones cannot be widely used since they have to be hidden,
and how this is done can greatly affect the quality of the sound obtained. Also, of course, no adjustment of position is possible to cater
for changes in the actors' positions.

The solution adopted in practice is to mount the microphone in a cradle at the end of a telescopic boom. This is mounted on a movable platform which stands about five feet above the floor. This can be man-handled into position and, by standing on the platform, the operator has a good view of the acting area. He can extend the boom and twist and tilt the cradle, so that the microphone is always in the best possible position. Cardioid microphones are generally used, so that any noise of the floor staff is heavily attenuated and, of course, the front-to-back ratio helps to obtain good direct sound even though the working distance is larger than desirable.

With many types of programme there is, of course, no objection to the microphone appearing in shot; panel shows and discussions are two examples. Here desk microphones, usually condenser cardioids of small size and unobtrusive appearance are used, or the people taking part wear halter type microphones round their necks. This makes the sound problem a lot easier.

Television Music

Returning to music programmes, we can appreciate that these can be of several distinct types, light music, jazz, opera, etc., and there are many problems involved in each of them. The difficulty of separating a vocalist from the accompanying orchestra, for example. In opera, it is often decided to use two studios, one for the orchestra and one for the singers. This introduces the need for special arrangements of loudspeakers, so that the singers can hear the orchestra and the conductor can hear the singers. This will show just some of the difficulties in televising certain types of musical programmes.

We have seen that the reverberation time in all television studios is low, and this gives a dry, unnatural sound when an orchestra is playing. If the viewers were the only consideration, this problem could be overcome to a great extent simply by adding artificial reverberation, or 'echo' as it is sometimes called, to the output of the studio.

Artificial Echo

Each microphone channel on the sound control equipment can feed a proportion of its output to a room, known as the echo room, which has highly reflective surfaces and is fitted with a loudspeaker and a microphone. The loudspeaker is fed with the combined outputs of the microphone channels, and the considerable reflections from the walls are picked up by the microphone, fed back to the sound control desk and mixed with the original studio output. The effect of this is to add

191

reverberation to the output of the studio. The proportion of each microphone's output fed to the echo room and the amount of the echo room output mixed in can be adjusted and controlled by the sound supervisor, to suit particular needs.

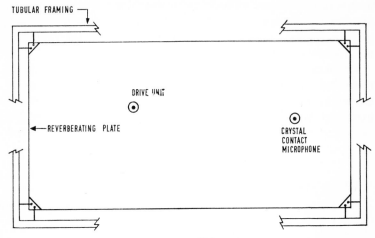

FIGURE 10.3

This shows the arrangements of the reverberating plate used for producing artificial reverberation. The thin steel sheet measures 1 × 2 metres and it is held under tension by a tubular steel frame. The drive unit produces transverse vibrations which are picked up by a contact microphone. Behind the plate is a sheet of porous material which can be moved to adjust the spacing and hence the reverberation time as required.

Limitations of Echo Rooms

Echo rooms suffer from the disadvantage that they are usually much smaller than the studio. Small rooms have strong colorations when they resonate and these colorations are audible. When an echo room is mixed with a studio, these colorations can be obvious to the listener. Since the acoustic qualities of a room or studio depend on the volume, shape, etc., the reverberant qualities of a small room are quite clearly not those of a large studio. The added reverberation does not have the quality which a lively enclosure of the dimensions of the original studio would have. Ideally, the echo room should be of large volume, but this is impossible from an economic point of view. Another important point is that the reverberation time of an echo room is fixed. In practice, however, to meet the demands of different studios and of different types of music, a range of reverberation times is required.

The Reverberation Plate

The need to alter the reverberation time has led to the development of various other methods of deriving artificial reverberation. One widely used in professional work is the reverberation plate, developed in Germany. A sheet of steel measuring 1 × 2 metres and 0·5 mm thick is suspended on a welded tubular steel frame. The plate is made to vibrate by a moving coil driving unit and these vibrations are picked up by a crystal contact microphone. These vibrations simulate the reflections occurring in a room (see Fig. 10.3).

The decay of the vibrations will depend on how much damping there is of the plate. This damping can be controlled, and hence so can the reverberation time, by adjusting the distance between the fixed steel sheet and a sheet of porous material. This is the same size

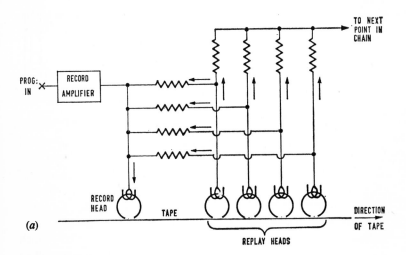

(a)

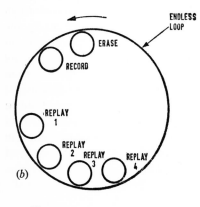

(b)

FIGURE 10.4

(a) Principle of the artificial reverberation machine.

(b) With this arrangement the medium can be used repeatedly.

as the steel sheet and is made of compressed glass fibre. By moving the porous sheet away from the steel sheet, the reverberation time can be increased, the range being of the order of 1 to 5 sec. The whole unit is housed in a wood-fibre box and this can be installed in any convenient room, which is reasonably quiet to avoid interference. The reverberation time can be controlled remotely from the sound control desk if required.

In broadcasting and recording organisations, the sound supervisor usually has at his disposal one or more echo rooms of different sizes, and some generator such as the plate, to provide a choice of reverberation times and types of reverberation.

The Performers' Point of View

While these techniques and devices improve very considerably the sound for the listener, they have no effect on the original acoustics of the studio. This is an important point, as orchestras are accustomed to playing in concert halls which have good acoustics from the players' point of view as well as that of the audience. In concert hall design, the architect must arrange the surfaces around the platform to reflect some sound back into the orchestra, so that the players hear themselves and each other. This leads to restrained playing and good ensemble.

But in the dead conditions of a television studio, these reflections are missing and this has a pronounced effect on the players, especially of the weaker instruments such as the strings. They tend to force the tone to hear themselves and this leads to poor quality, and ensemble playing is poor.

A straightforward and practical way of at least partially overcoming this difficulty is to surround the weaker sections of the orchestra with sound reflecting surfaces. This does help a great deal, but it cannot alter the reverberant qualities of the studio as a whole.

Ambiophony

A more elegant solution is now in use which makes the acoustics of the studio appear livelier by using electro-acoustic apparatus. If a loudspeaker were fixed to the wall of the studio, and fed with the amplified output of the orchestral microphones (care being taken to avoid howl-round), the loudspeaker output would simulate reflections from the wall of the studio. However, since the velocity of sound waves is comparatively low, the sound coming from the loudspeaker, fed by an electrical path, would occur earlier in time than a true acoustic reflection. This problem can be overcome by delaying the feed to the loudspeaker on a tape loop or by other means. Knowing the length of the acoustic path and the velocity of sound in air, the required time delay can easily be calculated.

This system is called ambiophony and was originally developed in Holland. Of course, a single loudspeaker and a single delay could not possibly imitate all the reflections in a normal enclosure and a large number of loudspeakers are used in practice, disposed around the

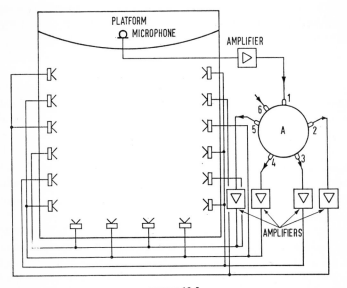

FIGURE 10.5

This layout diagram shows an installation for ambiophony. The rotating wheel *A* is coated on its edge with magnetic material. The outputs of the four spaced reproducing heads (2 to 5) are fed to the loudspeakers. The diffuse and delayed sound from these loudspeakers supplements the inadequate indirect sound in the auditorium. 1 and 6 are the record and erase heads respectively. (Courtesy Philips Technical Review.)

studio. These are fed from a magnetic recording system, having several reproducing heads, so spaced that their outputs are delayed to simulate the time it takes for the sound waves to reach various parts of the studio. Even with a large number of loudspeakers, the multiplicity of the reflections taking place in a studio cannot be reproduced. However, by taking care in the siting of the loudspeakers and in the adjustments of the delays of the feeds to them, very much improved acoustics can be produced with a beneficial effect on playing conditions for the orchestra.

A further development of this basic idea is to suggest that the studio is much bigger than it actually is. This is done by increasing the delay in the loudspeaker feeds so that they simulate an acoustical path of longer length. With a practical number of loudspeakers, there is obviously some limit to the increased size which can be suggested but, before that limit is reached, a useful improvement in acoustic conditions can be obtained (see Fig. 10.5).

The use of time delay has also had a very considerable effect on the design of high quality public address systems and these will be discussed later in the chapter.

'Foldback' Sound

We have already remarked that two studios are often used when televising opera, one for the orchestra and one for the singers. The output of the orchestral studio must be fed to the other studio for the singers, and here an obvious difficulty is how to confine the sound from the loudspeakers so that it is not picked up by the singers' microphone. The conductor too must hear the singers, and again it is important that the sound from his loudspeaker does not affect the orchestral microphones. (A speaker often has to be used in practice, since conducting wearing headphones is difficult.) It is only by using directional microphones and special loudspeaker systems that these difficulties have been overcome, to allow the broadcasting of live opera rather than relying on miming techniques.

Public Address Systems

There are two major faults with many public address (P.A.) systems, the tendency to instability or howl-round and the lack of intelligibility. In this section it is proposed to discuss how these faults can be avoided in simple installations of the sort required in amateur circles, and then see how the latest equipment and techniques are applied in professional high quality systems.

'Howl-round'

One of the most obvious faults in many P.A. systems is that they are too loud! Considerable improvement is often possible by turning down the loudspeaker volume. Not only does this make listening more tolerable to the public, but it reduces the risk of howl-round. This is caused, of course, by feedback of energy from the loudspeaker to the microphone, and it is important to remember that this energy can be returned by reflection from the walls of the hall.

Directional Microphones

On installing a simple P.A. system, therefore, one should note the directivity pattern of the microphone, the position of the loudspeakers and the reflectivity of the walls. A directional microphone is obviously much more useful than an omni-directional one, since it allows the loudspeaker to be placed in the dead areas of the microphone directivity pattern. Modern, robust moving coil cardioids, specially designed for P.A. work, were discussed in Chapter 5 and these have cut down very considerably the risk of howl-round.

In using a cardioid, the loudspeakers should be placed in front of the platform, raised above the heads of the audience and tilted so that the sound is directed on to the centre of the audience area. Not only does this put the sound where it is wanted, but the audience, due to their clothing, absorb sound energy and this helps to prevent feedback. With care, a simple system should not howl-round. But, if the loudspeaker output cannot be raised to a sufficient level without instability, the reflectivity of the walls should be considered. In many halls, the walls are of hard plaster giving no absorption at all. If possible, therefore, some heavy curtains or similar material should be hung over a fair amount of the wall surfaces. The rear wall should receive particular attention, as this is frequently a source of trouble.

The foregoing remarks apply with even more force, of course, if a microphone other than a cardioid is used. With a ribbon, for example, while it would be possible to place the loudspeakers to the sides of the microphone to take advantage of its 'dead' sides, one live face would be open to reflected sound from the hall. If several loudspeakers can be used, an improvement is sometimes possible by distributing these down the length of the hall. This will allow the level of each loudspeaker to be kept down, instead of trying to cover the entire hall with, say, two loudspeakers mounted near the platform.

Directional Loudspeakers

Of course it is a great help if the sound radiated from the loudspeakers can be confined over a narrow angle, to avoid the sound finding its way back to the microphones or on to the walls. The distribution of sound from ordinary loudspeakers mounted on a baffle or in a cabinet, is very dependent on frequency. At low frequencies it tends to be an omni-directional source, radiating equally in all directions; but, as the frequency rises, the radiation gets more and more directional until at high frequencies the sound is mainly confined to a forward beam. This variation of directivity with frequency is a draw-

197

back to good P.A. systems, since the wide distribution of low frequencies can cause trouble due to howl-round and, at the same time, the sharp beaming of the high frequencies is important as far as intelligibility is concerned. It is sometimes worth a trial to introduce some heavy bass cut, say below 200 c/s.

In Chapter 2 we noted that the intelligibility in speech is contained in the high frequencies. If the high frequencies from the loudspeaker are confined to a narrow beam, many members of the audience may have difficulty in making out the words being spoken. Of course, by using several loudspeakers, the high frequencies can be distributed more evenly over a wider area, and this is an additional argument for using more than one loudspeaker. We saw above how this could also reduce the possibility of a howl-round. It is also possible, of course, to get a wider distribution of high frequencies by using high frequency units, or tweeters.

With these refinements the distribution of sound from the speakers can be much improved. But there is a limit to what can be done, and this limit does fall short of what is required for an ideal system, free from instability and producing good intelligibility over the whole audience area.

This latter point has received considerable attention over the past few years, so that well-engineered systems are now used in places which at one time had notoriously bad listening conditions.

FIGURE 10.6
Column enclosure for simple line source system. The internal dimensions of the column are usually just large enough to accommodate the loudspeakers. (From *Loudspeakers*, by E. J. Jordan.)

Line-source Loudspeakers

A significant improvement is possible by using column or line-source loudspeakers. These consist of a series of loudspeakers mounted in line, or immediately adjacent to each other, so that they form a column facing in one direction. The loudspeakers are connected in phase, and their combined effect causes the distribution of sound to be confined to a broad horizontal beam with very little radiation in the vertical direction. When slung off the ground, and tilted so that the column is facing the audience, the sound is concentrated directly on to the audience and very little is radiated into the roof space. The immediate improvement is to increase the the direct sound reaching the audience and to reduce the reverberant sound, which has a beneficial effect on listening conditions. Further, the reduction in reverberant sound reduces the risk of howl-round.

The use of line-source loudspeakers has also made possible many of the techniques employed in broadcasting studios, especially television. When singers have to be able to hear the orchestra which is playing in a remote studio, for example, a line-source loudspeaker can be arranged vertically near the floor, so that the microphone picking up the singers is above the loudspeaker. The amount of orchestral sound picked up by the microphone is thus kept to a minimum.

Audience Microphones
When an audience is present in a broadcast studio, there are several difficulties. First of all, the audience have to be able to hear what is taking place near the microphone. In a large studio, and with the microphone techniques used by solo artistes and speakers, this is normally impossible unless loudspeakers are used to reinforce the sound. Secondly, in many types of programme it is necessary to pick up the reaction of the audience, and this means slinging a microphone over the heads of the audience with the consequent risk of howl-round. This can be reduced to a minimum by arranging the line-source loudspeakers used for public address so that the audience microphones are in the dead vertical area.

How Line-source Loudspeakers Work
The directivity of a line-source loudspeaker depends on its length, an obvious parallel with the line or 'rifle' microphone which was described in Chapter 5 (see p. 115). There we saw that the directivity is poor at low frequencies because the length must be limited in practice and the ratio of length to wavelength is therefore low at low frequencies. At high frequencies, when the length/wavelength ratio is large, the directivity is good. There is a similar effect with directional loudspeakers. At high frequencies the directivity is, in fact, too good and these are radiated in a narrow beam. Therefore, there can be a serious drop in high frequencies for small movements of the listener's head, with a consequent reduction in the intelligibility.

There are various ways of tackling this problem. One is to use two columns inside one column unit, a long one of say 9 ft using 6-in. units, and a short one of 3 ft using 3-in. units. The columns are placed side by side and fed through a network crossing over at about

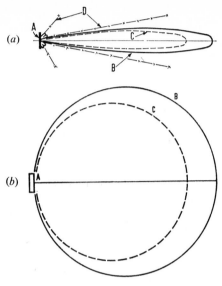

FIGURE 10.7
General form of contours of constant loudness for line-source loud-speakers.
(a) Vertical plane. B and C are two contours of equal loudness produced by a line source loudspeaker at A. D are directions of minimum radiation. Note the amplitude of the main lobes B and C compared with the amplitude of the side lobes.
(b) Horizontal. The contours B and C drawn for the horizontal plane show the wide coverage. (Courtesy Pamphonic Ltd.)

1,000 c/s. This tends to keep the length/wavelength ratio more constant and spreads the high frequencies.

Another method is to mount a rigid wooden panel in front of the speakers which has a slot ($1\frac{1}{2}$ in. wide is a quoted value) running down the entire length of the column. The distribution of sound through a slot depends in the ratio of the slot width to the wavelength. With a narrow slot, the distribution is more uniform over a wider frequency range than for a wide one. The effect of the narrow vertical slot is thus to keep the horizontal distribution nearly the same for high frequencies as for low frequencies.

Curving the column is another method. A quoted radius of curvature is twice the nominal column height, and the curving does give a wider distribution of high frequency sounds than a straight column using the same size of speakers.

Tapered Response

With a straight line-source loudspeaker, the directional pattern has a major lobe containing most of the power, but there are also side lobes, the ones immediately adjacent to the main lobe being about 15 dB down. These lobes can cause trouble due to their striking a distant surface and being reflected back on to the audience, and of course, they represent a waste of energy. These side lobes can be reduced by tapering the output power of each speaker so that maximum power is radiated by the central unit with a progressive reduction in power along the column towards each end. This tapering is simply done by controlling the power input to each loudspeaker, and it can practically double the attenuation of the first side lobe (Fig. 10.8).

With some line-source units, precautions are taken to reduce the radiation towards the back by using an open-back cabinet with a layer of packing material across the back of the loudspeaker. This material, which can be of mineral wool or cotton waste, acts as a delay path for the sound coming from the back of the speakers. It will be remembered that the radiation from the rear of the speaker is 180° out of phase with that from the front. If we now consider the sound from the front going round towards the rear of the column, after it has travelled half a wavelength it will be in phase with the radiation from the back, and the two will reinforce each other. If, however, the sound from the back of the speakers can be delayed by

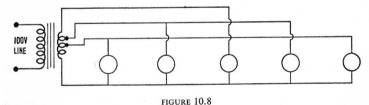

FIGURE 10.8
Method of power distribution used to reduce side lobes from a column loudspeaker. (From *Loudspeakers*, by E. J. Jordan.)

the same time as the front radiation takes to get to the rear of the column, the two radiations will still be 180° out of phase and they will cancel. The theory is somewhat similar to that used in the design of the phase shift microphone (see p. 94).

201

Time Delay

We have already mentioned the use of a time delay mechanism in the section on television studios, where it is used to improve the acoustics for the benefit of the musicians. This is really an advanced use of an idea which has been employed in high quality P.A. systems for some years to improve the naturalness of speech.

TABLE 10.1. *Line Source Loudspeakers*

	Single Unit Point-Source*	6 ft Type	8 ft 6 in. Type	11 ft Type
1. Total Vertical Main Beam Angle (4,000 c/s)	180°	37·6°	28°	22·6°
2. Relative Intensity at same Arbitrary Distance (4,000 c/s)	1	7·8	11·2	14·4
3. Relative Gain in dB (4,000 c/s)	0 dB	+8·9 dB	+10·5 dB	+11·6 dB
4. Relative Range (distance) for same sound level with equal power input (4,000 c/s)	1	2·8	3·35	3·8
5. Rated Maximum Peak Power	—	5 watts	10 watts	10 watts

* The point-source is assumed to have the same electro-acoustical power conversion efficiency as the columns.

TABLE 10.1

Line-source loudspeaker units are available in several different lengths and the table shows how various parameters vary for 6 ft, 8 ft 6 in., and 11 ft types. As discussed in the text the longer the unit the narrower is the beam width.

The choice of length depends on two main factors:

    (i) the acoustics of the hall

    (ii) the area to be covered with an adequate level.

If the hall has a long reverberation time a long column should be used so that the radiated energy is confined to as narrow a beam as possible. The second factor will decide the range of 'throw' required and here the longer unit will have the larger range since the rated input will be longer as can be seen in the table. (By courtesy of Pamphonic Reproducers Ltd.)

If someone is sitting near the platform in a hall, listening to a speaker, the direct path is short. There is no difficulty in understanding what is being said and the sound comes from where the listener's eyes tell him it ought to come. But if he is sitting well down the hall, and a P.A. system is being used, the direct sound will be weak and most of the sound striking his ears will come from the nearest P.A. loudspeaker.

We have seen earlier in the chapter how the velocities of the two paths—acoustical and electrical—are quite different. The sound from the loudspeaker will therefore arrive at the listener's ears before the

202

direct sound from the platform. The effect of this is to suggest to the listener that the source of sound is the loudspeaker, even though there is some direct sound and the listener can see the man speaking on the platform. This has a disturbing effect which robs the system of naturalness.

The manner in which human beings ascribe direction to sounds reaching him has already been touched upon in Chapter 3. How he ascribes direction between two sources was studied by Haas. He showed that, if sound reaches a listener from two loudspeakers, varying the times of arrival and the intensities of the sounds, could

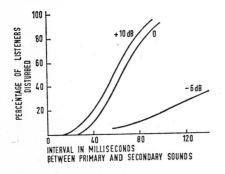

INTERVAL IN MILLISECONDS
BETWEEN PRIMARY AND SECONDARY SOUNDS

FIGURE 10.9
The effect of intensity difference and time difference on the disturbing nature of sound reaching listeners by two paths. If the secondary sound is greater than the primary one, the delay must be small for there to be a low percentage of listeners considering the secondary sound disturbing. In practice a 10% disturbance is considered a safe criterion. (From *Acoustics, Noise & Buildings*, Parkin & Humphreys. Faber & Faber.)

each alter the direction of the apparent source of the sound. And from this emerged the interesting fact that, within limits, it is the first sound to arrive which decides the direction, even though subsequent sounds may be stronger.

Returning to our P.A. system, we can now appreciate how this knowledge can be applied. If the feed to the loudspeaker can be delayed, so that the direct sound arrives first, the listener will ignore the loudspeaker even though the sound coming from it is much stronger than the direct sound. This allows the total sound intensity to be raised to an adequate level without causing the ear to ascribe a different direction to the sound from that judged by the eye and so avoiding the disturbing effect caused by the same sound arriving by different paths at different times.

Of course, if the intensity of the loudspeaker is increased above a certain level, or it is given too much delay, its presence becomes obvious. So there are practical limits. But before these limits are reached very useful and worthwhile improvements are obtained in practice.

It is found that the loudspeaker sound can be 10 dB above the direct sound, over a time-delay of between 5–25 milliseconds, without

203

an audience being aware of the loudspeaker. Reducing the loudspeaker intensity would allow an even longer delay.

Various methods are used in P.A. systems for delaying the speech, but essentially they consist of a magnetic recording arrangement with a series of spaced reproducing heads. Knowing the length to be covered, the delays required for loudspeakers placed at various points down the hall can be calculated and the relative distances of the reproducing heads adjusted accordingly.

Frequency Shift

We have seen how the room or hall can act as a coupling between a loudspeaker and a microphone, and lead to instability or howl-round. Earlier in the chapter the subject of eigentones or room resonances was discussed, and from this we can appreciate that a room has a frequency response which is characterised by a series of modes. If these modes are stimulated by the loudspeaker's output, the sound intensity in the room can be raised to a dangerous level.

A further refinement in P.A. work has been proposed by Schroeder of Bell Laboratories in which a frequency shift is given to the feed to the loudspeakers. The basic idea is that this will cause the loudspeaker radiation to be separated from the mode frequencies. The amount of shift given is two to three cycles, and this achieved by using a modulation technique so that the output signal is the same as the input signal, but is merely shifted slightly in frequency. It might be remarked that such an operation would alter the speech quality, but in fact the change is imperceptible. With this facility, the level of the loudspeaker output can be raised by several dB before howl-round occurs.

APPENDIX

ELEMENTARY TRIGONOMETRY (CIRCULAR MEASURE) AND DERIVATION OF A SINE WAVE

With reference to Fig. A.1, the trigonometrical ratios are defined as follows:

$$\text{Sine } \theta = \frac{AB}{OA}$$

$$\text{Cosine } \theta = \frac{OB}{OA}$$

$$\text{Tangent } \theta = \frac{AB}{OB}$$

The ratios are usually abbreviated and written as sin, cos and tan.

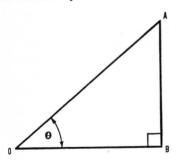

FIGURE A.1

For angles greater than 90° it would be possible to have the same ratios for different angles, and a sign convention is used to avoid ambiguity.

Using the triangle OAB, we have defined sin, cos and tan. However, if we rotated OA until $AB' = AB$ as in Fig. A.2, the ratios of the sides would be the same and yet θ is obviously not equal to θ'.

The line denoted by OA is always considered positive, but the signs of the other two sides depends on the direction in which they are measured from O. For example, OB is positive but OB' is negative. Similarly with OY and OY'. Dividing the 360° into four quadrants, we can see that in the first quadrant all the ratios are

205

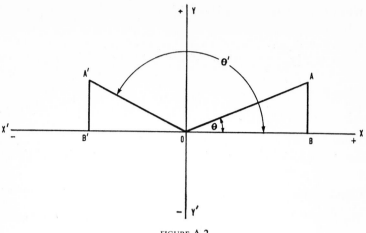

FIGURE A.2

positive since directions along OX and OY are used. In the second quadrant, the sin is positive since AB is in the Y direction; the cosine will be negative since OB' is in the X' direction and from this it follows that the tangent will also be negative.

A diagrammatic way of showing the quadrants in which the ratios are *positive* is shown in Fig. A.3.

Referring again to Fig. A.2, and considering θ to be a small angle, AB must be small and OB must be nearly equal to OA. Put more precisely, as θ tends to zero OB and OA tend to equality.

It follows therefore that:
When $\theta = 0$,

$$\text{Sin } \theta = 0, \cos \theta = 1 \text{ and } \tan \theta = 0.$$

FIGURE A.3

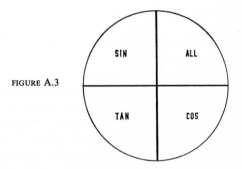

When θ tends to 90°, OA and AB tend to equality, and OB tends to 0.

It follows therefore that:

When $\theta = 90°$

$$\text{Sin } \theta = 1, \cos \theta = 0 \text{ and } \tan \theta = \infty$$

When $\theta = 180°$

$$\text{Sin } \theta = 0, \cos \theta = -1 \text{ and } \tan \theta = 0$$

When $\theta = 220°$

$$\text{Sin } \theta = -1, \cos \theta = 0 \text{ and } \tan \theta = \infty$$

Circular Measure

The ratio $\dfrac{\text{circumference}}{\text{radius}}$ is a constant for all circles and its value is 6·2831 . . ., which is usually written as 2π.

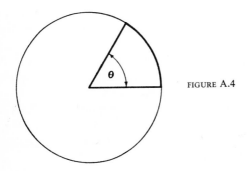

FIGURE A.4

An arc is a length measured on the circumference of a circle, and for any given value of included angle the ratio of this arc to the radius of the circle is a constant for all circles. The magnitude of this constant will depend on the angle between the two radii forming the arc (Fig. A.4).

The ratio arc/radius is called the circular measure of the angle. The unit is the *radian*, which is the angle formed when the arc and radius are equal. An angle of 2 radians means that the arc is twice the length of the radius, and so on. From this we can get the relation between angles measured in degrees and radians.

The ratio for a complete circle $\dfrac{\text{circumference}}{\text{radius}} = 2\pi$, and this corresponds to an angle of 360°.

207

That is 2π radians $= 360°$

and so π radians $= 180°$

$$\frac{\pi}{2} \text{ radians} = 90°$$

Derivation of a Sine Wave

Referring to Fig. A.5, let the radius OA be rotating in an anti-clockwise direction at an angular velocity of ω radians/sec. After a time, t, the angle $\widehat{AOB}$ will be ωt.

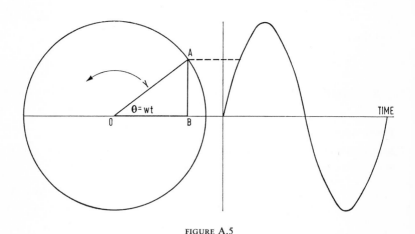

FIGURE A.5

$$\text{Now } \sin \phi = \sin \omega t = \frac{AB}{OA}$$

or $AB = OA \sin \omega t$

$$\therefore \quad y = Y \sin \omega t$$

whereby y is the value of AB at time t.

Inserting values for ωt:

When $\omega t = 90°$ $\sin \theta = 1$ and $y = Y$
When $\omega t = 180°$ $\sin \theta = 0$ and $y = O$
When $\omega t = 270°$ $\sin \theta = 1$ and $y = -Y$
When $\omega t = 360°$ $\sin \theta = 0$ and $y = O$

$y = Y \sin wt$ is thus an equation for which all the values can be plotted or represented by a graph. This is shown in Fig. A.5, and is called a sine curve.

This can be used to represent any quantity which varies sinusoidally with time, e.g. pressure in a sound wave, displacement of air particles, alternating voltages, etc.

The periodic time, T, is the time taken for one complete revolution.

i.e.
$$T = \frac{2\pi}{\omega} \text{ sec.}$$

Since $f = \dfrac{1}{T}$

$$\omega = 2\pi f$$

and the equation can be rewritten as:

$$y = Y \sin 2\pi f t.$$

Y is termed the amplitude.

Phase Angle

Suppose that when ϕ is 90°, and y a maximum, another radius Y' starts to rotate. Fig. A.6 shows how the two curves are related with

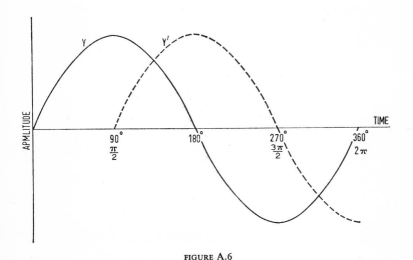

FIGURE A.6

respect to time, and there is said to be a phase angle of 90° or $\frac{\pi}{2}$ radians between them.

Mathematically this is represented by writing y' as

$$y' = Y \sin \left(\omega t - \frac{\pi}{2} \right) [Y = Y']$$

The negative sign indicates a lag in phase; the maximum value of y' is reached a quarter of a period after y in this example.

A positive sign indicates a lead in phase. The choice of sign depends on the choice of reference radius; y' lags on y OR y leads on y'.

Standing Waves

Complete and partial reflection. Measurement of absorption coefficient by standing wave method

$$A \cdot \xleftarrow{\hspace{1cm} l \hspace{1cm}} \cdot B$$

Let the pressure at A be $P \sin \omega t$, and assume that the wave is travelling from A to B. The wave will arrive at B after a time interval $\frac{l}{c}$ where c is the velocity of the wave. The pressure at B must lag on that at A; that is there is a phase difference between the two pressures. This difference is shown by denoting the pressure at B by:

$$P' = P \sin \omega \left(t - \frac{l}{c} \right)$$

This is usually written as:

$$\underline{P \sin (\omega t - kl)} \text{ where } k = \frac{2\pi}{\lambda}$$

The negative sign indicates movements from left to right. Movements in the opposite direction are indicated by a positive sign, as in $P \sin (\omega t + kl)$.

We now consider an incident wave $P \sin (\omega t - kl)$ striking a rigid, non-absorbent surface. The reflected wave will be $P \sin (\omega t + kl)$. The pressure in the combined wave will be:

$$\begin{aligned} P'' &= P \sin (\omega t - kl) + P \sin (\omega t + kl) \\ &= P [\sin (\omega t - kl) + \sin (\omega t + kl)] \\ &= P [2 \sin \omega t \cos kl] \end{aligned}$$

(Since Sin $(A + B) +$ Sin $(A - B) = 2$ sin A cos B)

$$= 2P \text{ sin } \omega t \text{ cos } kl$$

This result shows that the pressure varies with time and distance in the manner that was deduced graphically in Chapter 1 (see p. 28).

Perfect Reflection

By taking the rigid surface as our reference point, we can also find how the pressure varies with distance.

(a) At the reflecting surface, $l = 0$, cos $kl = 1 \cdot 0$ and the pressure varies sinusoidally with time, reaching a maximum value which is twice the amplitude of the incident wave. This is an *ANTI-NODE*.

(b) Quarter of a wavelength from the surface, $l = \dfrac{\lambda}{4}$ and cos $kl = 0$.

The pressure at this point is zero and remains at zero at all times. This is a *NODE*.

(c) Half a wavelength from the surface, $l = \dfrac{\lambda}{2}$ and cos $kl = -1$.

This is again an anti-node, but the pressure is reversed in phase compared with that at the reflecting surface.

Partial Reflection

We now consider a reflecting surface which absorbs some of the incident energy so that the reflected wave will be smaller in amplitude than the incident wave. Let the fraction of sound pressure reflected be m.

$$\text{The incident wave} = P \text{ sin } (\omega t - kl)$$

$$\text{The reflected wave} = mP \text{ sin } (\omega t + kl)$$

$$\text{The combined wave} = P \text{ sin } \omega t \text{ cos } kl - P \text{ cos } \omega t \text{ sin } kl$$

$$+ mP \text{ sin } \omega t \text{ cos } kl + mP \text{ cos } \omega t \text{ sin } kl$$

$$= (P + mP) \text{ sin } \omega t \text{ cos } kl - (P - mP)$$
$$\text{cos } \omega t \text{ sin } kl$$

$$= P(1 + m) \text{ sin } \omega t \text{ cos } kl - P(1 - m)$$
$$\text{cos } \omega t \text{ sin } kl$$

The wave consists of two stationary waves having a phase difference of $\dfrac{\pi}{2}$ and having amplitudes of $P(1 + m)$ and $P(1 - m)$.

(a) At the reflecting surface, $l = 0$, cos $kl = 1$ and sin $kl = 0$. The pressure is varying sinusoidally with time and has a maximum value $P(1 + m)$.

(b) Quarter of a wavelength from the surface, $l = \dfrac{\lambda}{4}$, cos $kl = 0$ and sin $kl = 1$. The pressure is again varying sinusoidally with time and has a maximum value $P(1 - m)$.

We see that instead of having a pressure equal to twice the incident pressure at the anti-node, and complete cancellation at the nodal point, there are now maxima and minima.

The relative values of the maxima and minima depend on 'm' and hence on the absorption of the reflecting surface. By measuring the maxima and minima, we can obtain the absorption co-efficient.

This absorption co-efficient (α) is defined as the ratio of the absorbed energy to the incident energy, i.e.

$$\alpha = \frac{\text{Absorbed Energy}}{\text{Incident Energy}}$$

$$= \frac{\text{Incident Energy} - \text{Reflected Energy}}{\text{Incident Energy}}$$

$$= 1 - \frac{\text{Reflected Energy}}{\text{Incident Energy}}.$$

But the energy is proportional the square of the pressure.

$$\therefore \quad \alpha = 1 - \frac{(mP)^2}{(P)}$$

$$= 1 - m^2$$

Let the pressure measured at a maximum be $A1$ and that at a minimum be $A2$.

Then

$$\frac{A1}{A2} = \frac{P(1 + m)}{P(1 - m)}$$

$$m = \frac{A1 - A2}{A1 + A2}$$

and

$$\alpha = 1 - \left\{\frac{A1 - A2}{A1 + A2}\right\}^2$$

This practical method of determining the absorption co-efficient of samples of materials uses a pipe some six feet in length and one foot in diameter. It is of earthenware and is closed at one end by the sample under test. The other is coupled to a loudspeaker which acts as the source of sound. A probe microphone fitted on the end of a sliding tube is passed into the pipe, through a small hole in the sample, and used to measure the maxima and minima pressures.

There is one fundamental objection to this method. The sound

212

waves strike the sample at right angles, and this is not necessarily the case when the material is used to treat a room.

A more accurate method is to use a reverberation chamber and note the effect on the reverberation time of various materials. This is more representative of what will happen in practice, but the measurement is time consuming and relatively more expensive. The tube method just described gives quick, easy evaluations. Published data on absorption co-efficients usually indicate by which method they have been measured.

Bels and Decibels

The Bel is a unit on a logarithmic scale expressing the ratio of two intensities or two powers. The number of Bels, N, is the logarithm (to the base 10) of the ratio.

$$\text{i.e. } N = \log_{10} \frac{\text{Intensity}_1}{\text{Intensity}_2} = \log_{10} \frac{I_1}{I_2}$$

The decibel is one-tenth of a Bel. If the number of decibels is n:

$$n = 10 \log_{10} \frac{I_1}{I_2}$$

Since intensity is proportional to the square of the pressure, the decibel equation can be rewritten as

$$n = 10 \log \frac{\text{Pressure}^2_1}{\text{Pressure}^2_2}$$

$$= 20 \log \frac{P_1}{P_2}$$

The above manipulation depends on the factors of proportionality cancelling out. This is true when plane progressive waves are being measured, but for other types the intensity ratios are not necessarily directly proportional to the square of the pressure ratios. However, since the measurement in much acoustic work is done by microphones which produce output voltages proportional to pressure, it is convenient to measure the pressures in a sound field and insert the values directly into the decibel formula.

In electrical communication work, the same ideas apply. Strictly speaking, decibels depend on the ratio of two powers.

i.e. if P_1 is the output power and P_2 the input power,

$$n = 10 \log_{10} \frac{P_1}{P_2}$$

If $P_1 < P_2$ then there is a loss; if $P_2 < P_1$ there is a gain.

In practice it is more convenient to measure voltage or current than power.

Since $P = \dfrac{E^2}{R}$ or I^2R (from Ohm's Law) it is possible to derive alternative expressions for the decibel ratio.

i.e.
$$n = 10 \log_{10} \frac{\dfrac{E_1{}^2}{R_1}}{\dfrac{E_2{}^2}{R_2}} \quad \text{or} \quad 10 \log_{10} \frac{I_1{}^2 R_1}{I_2{}^2 R_2}$$

Provided $R_1 = R_2$, i.e. the measurements are made or related to points in the circuit where the impedances are equal,

$$n = 20 \log_{10} \frac{V_1}{V_2} \quad \text{or} \quad 20 \log \frac{I_1}{I_2}$$

Therefore, if the resistances of both circuits are the same, one can use either voltage or current and get the same result as when power is considered.

In practice, however, the voltage or current ratios are sometimes used although the resistances or impedances of the two circuits are not equal. Therefore in using the decibel in acoustics or electrical communications, care must be taken to specify what quantity is being used and under what conditions it is being measured.

Reference levels are used in both acoustics and communications. Zero acoustical level has an intensity of 10^{-16} watts/sq. cm which corresponds to a pressure of 1 dyne/sq. cm. Zero level in communications is 1 milliwatt.

There are other reference levels in use for various measurements. It is important to state whether one of these is being used.

Doppler Effect

This is the effect on the pitch of a sound when there is relative movement of the source and the listener.

Let:
    frequency of source $= f_s$
    frequency at observation point $= f_o$
    velocity of sound $= c$
    velocity of listener $= v_1$
    velocity of source $= v_s$

Then
$$f_o = \frac{c \pm v_1}{c \pm v_s} f_s$$

The choice of sign depends on the direction of the movement, e.g. if the source is moving towards the listener who is stationary, the expression becomes

$$f_0 = \frac{c}{c - v_s} f_s$$

If the source is moving away from a stationary listener,

$$f_0 = \frac{c}{c + v_s} f_s$$

PAPERS

AXON, P. E., GILFORD, C. L. S., and SHORTER, D. E. L. 'Artificial Reverberation', *Proc. I.E.E.*, Vol. 102, Part B, 1955.

KUHL, W. 'The Acoustical and Technological Properties of the Reverberation Plate'—European Broadcasting Review Part A, Technical No. 49, May 1958.

PARKIN, P. H. and SCHOLES, W. E. 'Recent Developments in Speech Reinforcement Systems', *Wireless World*, Vol. LVII, No. 2, February 1951.

KLEIS, D. 'Modern Acoustical Engineering—I. General Principles', *Philips Technical Review*, Vol. 20, No. 11, 1958/59.

'The Broadcasting of Music in Television', BBC Monograph No. 40, February 1962.

SOMERVILLE, T. and GILFORD, C. L. S. 'Acoustics of Concert Halls', *Proc. I.E.E.*, Vol. 104B, No. 85, 1957.

GILFORD, C. L. S. 'The Acoustics Design of Talks Studios and Listening Rooms', *Proc. I.E.E.*, Vol. 106, Part B, No. 27, 1959.

SHORTER, D. E. L. 'Operational Research on Microphone and Studio Techniques in Stereophony', BBC Monograph No. 38, September 1961.

SHORTER, D. E. L. and PHILLIPS, G. J. 'A Summary of the Present Position of Stereophonic Broadcasting', BBC Monograph No. 29, April 1960.

REFERENCES

PARKIN, P. H., and HUMPHREYS, H. R. *Acoustics, Noise and Buildings* (Faber and Faber Ltd.). This book is the work of a scientist and an architect and gives much practical information on the solution of acoustic problems as well as a useful review of the basic physical concepts.

GAYFORD, M. L. *Acoustical Techniques and Transducers* (Macdonald and Evans Ltd.). This book covers loudspeakers, microphones, gramophone pickups, room acoustics, vibration measurements and stereophonic reproduction. There is much that is easily accessible to the Hi-Fi enthusiast.

OLSON, H. F. *Elements of Acoustical Engineering* (D. Van Nostrand Co. Inc.). Although written primarily for engineers, the amateur would find a great deal of interesting reading in this book which covers a very wide range of subjects.

BURRELL HADDEN, H. *High Quality Sound Production & Reproduction* (Iliffe Books Ltd.). This book was produced in the Central Programme Operations Department of the BBC for their own personnel, both technical and non-technical, to enable them to obtain the best results from studio equipment. There is much to interest the amateur, as it deals with microphones and their placement and loudspeakers.

217

VAN SANTEN, G. W. *Mechanical Vibration 2nd Edition* (Phillips Technical Library). This book reviews the elementary theory of mechanical vibrations and although it uses advanced mathematics in places, there is much interesting material.

ROBERTSON, A. E. *Microphones* (Iliffe Books Ltd.). This covers the basic operating principles of all modern microphones at a level suitable for those with average mathematical ability. It is primarily written for BBC Staff and contains much useful and interesting information as to how microphones work.

BERANEK, LEO. *Music, Acoustics and Architecture* (John Wiley and Sons Inc.). Although this is an expensive book, it is well worth reading for those interested in music. Dr. Beranek travelled on five continents listening to music and making measurements in over sixty halls. He is an international authority on acoustics.

OLSON, H. F. *Musical Engineering* (McGraw-Hill Book Co. Inc.). This book was written for musicians, engineers and enthusiasts interested in speech, music, musical instruments, acoustics, sound reproduction and hearing. The major portion of the book employs simple physical explanations, illustrations and descriptions which can be read without any special training in music, engineering, physics or mathematics.

WOOD, A. *The Physics of Music* (Methuen & Co. Ltd.). This gives an introduction to the very interesting borderland between physics and music. Starting with a review of the physics of sound, it goes on to deal with hearing, musical quality and musical instruments.

JEANS, Sir JAMES. *Science and Music* (Cambridge Press). Although written over twenty years ago this is still an excellent book. The author's aim was to convey precise information in a simple non-technical way. In doing this he wrote a book which has always enjoyed a very high reputation.

GLOSSARY OF TERMS

ABSORPTION CO-EFFICIENT – Measure of the efficiency of an absorbing material or method. It is the ratio of the sound energy absorbed to the incident sound energy.

ACOUSTICS – The science of sound.

ACOUSTIC DOUBLET – A double sound source, e.g. the front and rear surfaces of a diaphragm or loudspeaker cone.

ACOUSTIC FEEDBACK – The transfer of physical vibrations from a loudspeaker to other apparatus in the reproducing chain in such a way that spurious electrical signals are fed back to the loudspeaker.

ACOUSTIC TREATMENT – The application of absorbent or reflecting material to the walls, floor, ceiling, of a room to alter its acoustic properties.

AMPLITUDE – The peak or maximum value of a vibration or wave motion.

ANECHOIC – Without echo. An anechoic chamber is a chamber or room where walls are lined with a material which completely absorbs sound.

ANTI-NODE – A point, line or surface in a standing wave system at which the amplitude is a maximum, e.g. an anti-node of pressure, an anti-node of particle velocity.

AUDIO-FREQUENCY – Rate of oscillation corresponding to that of sound audible to the human ear, i.e. from 16 to 16,000 cycles per second approx.

BAFFLE – General expression for wall, board or enclosure carrying the loudspeaker. The purpose of the baffle is primarily to separate the front and back radiations from the cone or diaphragm which would otherwise cancel each other.

BASS – The name given to the lower frequencies in the audio range.

BASSY – Sound reproduction that over-emphasises low-frequency tones.

BEATS – Audible difference tones produced when two signals of nearly the same frequency are sounded together.

BIMORPH – Two thin slabs of crystal cemented together.

CAPACITOR – A device for storing an electric charge, analogous to the compliance of a spring which may be used to store mechanical energy.

CARDIOID MICROPHONE – Class of microphone having a heart-shaped directivity pattern in the horizontal plane.

COINCIDENT – Refers to microphone arrangements in stereophony. Two microphones are said to be coincident if they are placed immediately adjacent to each other so that any differences in the times of arrival of the sound are negligible.

COMPATIBILITY – A stereophonic system which provides a signal capable of giving satisfactory results on monophonic equipment.

COMPRESSION – Part of a sound wave where the pressure is above the standing atmospheric pressure.

CONSONANCE – A combination of two tones which is generally accepted as producing a satisfying effect.

CRYSTAL – In most pickups a bimorph of Rochelle Salt—two crystal slices cemented together with a conducting cement.

CRYSTAL MICROPHONE – Type of microphone in which a fluctuating voltage is generated by applying pressure to a slab of material such as Rochelle Salt or Barium Titanate.

c/s – cycles per second.

CYCLE – One complete series of changes in a periodically varying quantity.

DEAD ACOUSTIC – The dull acoustic effect of an enclosed space with little reverberation.

DECIBEL – A unit ratio of power, voltage or current.

$$N\ (\mathrm{dB}) = 10 \log_{10} \frac{P_2}{P_1},$$

219

where P_1 and P_2 are two amounts of power. When the powers are dissipated in equal impedances the voltage or current ratios may be determined as follows:

$$N \text{ (dB)} = 20 \log_{10} \frac{E_2}{E_1} \text{ or } 20 \log_{10} \frac{I_2}{I_1}$$

where E_1 and E_2 are the respective voltages and I_1 and I_2 are the respective currents.

DIFFRACTION – Bending of waves round solid objects. Low frequency sound waves bend round the edge of the loudspeaker enclosure and produce some re-radiation from the edges which may provide irregularities in the frequency response.

DIRECTIVITY PATTERN – Graph showing the response of a piece of equipment such as a microphone at all angles in a given plane—sometimes called a polar diagram.

DISPLACEMENT – Distance of a particle from its normal or mean position.

DISSONANCE – A combination of two tones which does not form a consonance.

DYNAMIC RANGE – The ratio (in phons) between the softest sound and the loudest sound in a live performance. The ratio (in dB) of the largest signal a given system can handle without distortion to the smallest signal that it can reproduce successfully (i.e. without the inherent background noise masking it).

DYNE – The unit of force in the metric system (C.G.S.). That force which will give to a moving mass of 1 gm an acceleration of 1 cm. per sec. per sec.

ELECTROSTATIC MICROPHONE – Class of microphone in which a fluctuating electric current is produced by the movement of a diaphragm relative to a rigid plate.

FREQUENCY – The number of vibrations of a body or particle completed in a second.

FUNDAMENTAL – The lowest frequency component of a complex wave.

HARMONIC – A component of a complex wave which is an integral multiple of the fundamental, e.g. twice the fundamental, three times the fundamental.

HELMHOLTZ RESONATOR – An acoustic resonator comprising a cavity having an aperture open to free air. This aperture may be fitted with a duct or tunnel either internally or externally.

INCIDENT WAVE – Wave travelling towards a reflecting or refracting surface.

INTENSITY – The rate of flow of sound energy per unit of area normal to the direction of propagation.

INTERVAL – In music, the pitch interval between two notes of a scale.

LEVEL – The strength of a continuous signal used for test purposes. Measured by comparison with a standard reference level, which is usually a power of 1 mW.

LINE SOURCE – Ideally an acoustic radiator having the dimension of length only. An array of loudspeakers arranged in a line.

LIVE ACOUSTIC – The bright acoustic effect of a room with considerable reverberation.

LOBES – Variation from a minimum value, through a maximum and back to minimum in the polar response of a radiator.

LONGITUDINAL – Of wave motion in which the particles of the medium vibrate along the line of propagation of the wave.

LOUDNESS – The magnitude of the auditory sensation produced by a sound. The unit used in modern work is the sone. The loudness scale is linear in that, for example, 2 sones sound twice as loud as 1 sone. More generally, x sones sound x times as loud as 1 sone. The sone and phon are related as follows:

(i) 1 sone is the loudness of a sound whose loudness level is 40 phons.

(ii) Doubling the loudness is equivalent to increasing the phon value by 10, e.g. raising the loudness from 1 to 2 sones raises the loudness level from 40 to 50 phons.

LOUDNESS LEVEL – The loudness level of a sound is in phons when it is judged to be equal in loudness to a pure tone of 1,000 c/s from a plane progressive wave. The pressure level of the 1,000 c/s tone being in dB above 2×10^{-4} dyn/cm^2, the standard reference pressure.

LOUDSPEAKER – A system for converting electrical energy into sound energy.

MASKING – The effect on the hearing of one sound when another sound is present. The shift of the intensity threshold of audibility of the first sound due to the presence of the second one.

MICROPHONE – A kind of electric ear which is finely sensitive to sound impinging on its diaphragm. It is used to convert acoustic energy into electrical energy, i.e. it is a vibration generator.

MODE – Manner or fashion. Resonant modes – manners in which a complex system resonates.

MODULATION – Moving from one key to another.

NODE – A point, line or surface in a standing wave system at which the amplitude is zero, e.g. a node of pressure, a node of particle velocity.

OCTAVE – A pitch interval of 2 : 1.

OVERTONE – A component of a complex wave which may or may not be an integral multiple of the fundamental.

PARTIAL – Any component of a complex sound; overtones and harmonics are partials.

PERIOD OR PERIODIC TIME – Time taken by an alternating quantity to perform one cycle.

PHASE – Two waves are *in phase* if they are always exactly in step.

PHASE ANGLE – A measurement of the phase difference between two waves.

PHON – The unit of equivalent loudness used in measuring loudness level.

PIEZO-ELECTRIC – The phenomena by which certain crystals (Rochelle Salt, quartz, Tourmaline, etc.) expand along one axis and contract along another, when subjected to an electrostatic field or, when subjected to twisting and bending, certain crystals develop a potential difference between opposite faces.

PITCH – That subjective quality of a sound which determines its position in the musical scale.

POLAR RESPONSE – A plot of the variation in radiated energy with angle relative to the axis of the radiator.

PRESSURE GRADIENT OPERATION – Where the sound has access to both sides of the diaphragm.

PRESSURE OPERATION – Where the sound wave has access to one side of the diaphragm only.

RAREFACTION – Part of a sound wave where the pressure is below the pressure due to standing atmospheric pressure.

REFLECTION OF SOUND WAVE – A return of energy due to the wave striking some discontinuity in its supporting medium.

REFRACTION – Bending of an incident wave due to its striking, at an angle, a change of medium.

RESONANCE – The effect when the vibrations of a body reach maximum amplitude on being caused to vibrate by a force having a particular frequency.

REVERBERATION – The prolongation of sound in a room due to reflections from the walls, ceiling and floor.

REVERBERATION TIME—— The time taken for the sound intensity in a room to fall from its steady state by 60 dB (i.e. to a millionth).

SABINE – Unit of sound absorption. One sabine corresponds to that absorption which would be pro-

221

duced by 1 sq. ft of an infinitely absorbing surface such as an open window. The average person has an absorption of 4·2 sabines.

SINE WAVE – The wave shape of a pure note.

SOUND – Any change in the ambient air pressure which has an amplitude and frequency content within the response range of the human ear.

SOUND STAGE – Refers to stereophony. This is the region in which the sound images appear. For each source of sound in the studio there will be a corresponding point from which the reproduced sound appears to come in the region between the two spaced loudspeakers. These points are termed images.

SPACED MICROPHONES – Refers to microphone arrangements in stereophony. Microphones are said to be spaced when the stereophonic effects are produced by differences in time of arrival at the microphones.

STANDING or STATIONARY WAVE SYSTEM – An interference pattern characterised by nodes and antinodes.

STEREOPHONIC – A sound recording/ reproducing system using a number of microphones and loudspeakers to achieve a spatial distribution of the reproduced sound.

TIMBRE – The tone colour which enables a listener to recognise the difference between two sounds having the same loudness and pitch.

TRANSDUCER – A device which converts power of one type into power of another, e.g. a loudspeaker.

TRANSIENT – The effect on a vibrating system when there is a sudden change of conditions, and which persists for only a short time after the change has occurred.

TRANSVERSE – Of a wave motion in which the particles vibrate at right angles to the direction of propagation.

VELOCITY OF SOUND IN AIR – $3·44 \times 10^4$ centimetres per second.

WATT – Unit of electrical power.

WAVEFORM – The configuration of a sound wave.

WAVELENGTH – The minimum distance between two points of a wave which are in phase.

WAVELENGTH (SOUND WAVE) – The distance travelled by the sound in the time of one complete vibration. Thus $\lambda = \dfrac{c}{f}$, where c is the velocity of sound in air (1,087 ft per sec at 0° C, rising 2 ft per sec per ° C increase in temperature) and f is the frequency.

INDEX